DYNAMICS IN ORGANIC MATTER PROCESSING, ECOSYSTEM METABOLISM AND TROPHIC SOURCES FOR CONSUMERS IN THE MARA RIVER, KENYA

Frank Onderi Masese

Thesis committee

Promotor
Prof. Dr Kenneth A. Irvine
Professor of Aquatic Ecology and Water Quality Management
UNESCO-IHE Institute for Water Education
Delft, The Netherlands

Co-promotors
Prof. Dr Michael E. McClain
Professor of Ecohydrology
UNESCO-IHE Institute for Water Education
Delft, The Netherlands

Dr Gretchen M. Gettel
Senior Lecturer in Aquatic Biogeochemistry
UNESCO-IHE Institute for Water Education
Delft, The Netherlands

Other members
Prof. Dr Wolf Mooij, Wageningen University
Prof. Dr Robert Naiman, University of Washington, USA
Prof. Dr Steve Ormerod, Cardiff University, UK
Dr Martin Pusch, Leibniz-Institute of Freshwater Ecology and Inland Fisheries, Germany

This research was conducted under the auspices of the SENSE Research School for Socio-Economic and Natural Sciences of the Environment

DYNAMICS IN ORGANIC MATTER PROCESSING, ECOSYSTEM METABOLISM AND TROPHIC SOURCES FOR CONSUMERS IN THE MARA RIVER, KENYA

Thesis

submitted in fulfilment of the requirements of
the Academic Board of Wageningen University and
the Academic Board of the UNESCO-IHE Institute for Water Education
for the degree of doctor
to be defended in public
on Wednesday, 11 March 2015 at 3 p.m.
in Delft, the Netherlands

by

FRANK ONDERI MASESE

Born in Kisii, Kenya

CRC Press
Taylor & Francis Group
Boca Raton London New York

CRC Press is an imprint of the
Taylor & Francis Group, an **informa** business

A BALKEMA BOOK

First issued in hardback 2018

CRC Press/Balkema is an imprint of the Taylor & Francis Group, an informa business

© 2015, Frank Onderi Masese

Published by:
CRC Press/Balkema
PO Box 11320, 2301 EH Leiden, The Netherlands
e-mail: Pub.NL@taylorandfrancis.com
www.crcpress.com – www.taylorandfrancis.com

ISBN 13: 978-1-138-37331-0 (hbk)
ISBN 13: 978-1-138-02815-9 (pbk)

Dedication

To the loving memory of my mother, the late Miriam Gesare Masese

Acknowledgements

Leaping across streams every week-day for school nearly three decades ago, my early days in formal education were intimate with nature. In these streams, I and my childhood friends fished with baskets and hook and line, and I was always intrigued by the diversity of aquatic macrofauna. Even though arduous and winding, my academic journey has gone full circle and taken me back to streams. The success has however been achieved with a dose of appreciation, and I thank God for His providence and the many people to whom I owe so much.

This dissertation has been made possible by the guidance and support of my promoters, mentors, colleagues, financial assistance from the Dutch Ministry of Foreign Affairs and support and encouragement from my wife. Firstly, I would like to thank my promoter and mentor Prof. Michael E. McClain (UNESCO-IHE, The Netherlands), for his support, mentoring and advice throughout the entire PhD program, as well as for giving me the opportunity to work on this project. I would also like to thank Michael for always being approachable and happy to discuss ideas and challenges, and for providing timely and constructive reviews of manuscripts and draft chapters. I would like to thank my promoter Prof. Kenneth Irvine and supervisor Dr Gretchen Gettel (UNESCO-IHE, The Netherlands), for their helpful advice, comments and reviews of manuscripts and dissertation chapters and the wealth of experience and different approaches they brought to my advisory board. I appreciate the guidance and support offered by Dr Nzula Kitaka (Associate Professor) and Dr Julius Kipkemboi (Associate Professor) of Egerton University, Kenya. I also appreciate the input offered by Prof. Steven Bouillon (KU Leuven, Belgium) and Dr. Katya Abrantes (James Cook University, Australia) for their advice and assistance during analysis of stable isotopes data.

I acknowledge the assistance offered by many people during field work and in the laboratory which made the compilation of this dissertation a success. To Kimutai Kitur and Robertson Marindany, thanks for spending days in the field collecting water quality data and samples, fish, bugs, and plants from the crocodile and hippo infested Mara River. Thanks to Dr Phillip O. Raburu (Associate Professor, University of Eldoret) for the support, guidance and logistical assistance during field work. I acknowledge Dr William Ojwang and Chrisphine Nyamweya (KMFRI, Kisumu) and their team for assistance during fish sampling and gut content analysis. Special thanks to Lubanga Lunaligo and David Namwaya (University of Eldoret) for assistance during lab work; I cannot forget to mention the commitment with which they helped open guts of bugs and sort litter. I thank Zita Kelemen (KU Leuven), Amanda Subalusky and Dr Glendon Hunsinger (Yale University, CT, USA) for stable isotope analysis. Thank you Amanda for the many discussions we had by the river and the encouraging words. I appreciate Dr Fidelis Kilonzo (Kenyatta University, Kenya and UNESCO-IHE), Dr Paolo Paron and Veronica Minaya (UNESCO-IHE) for their assistance with the maps used in this dissertation. I would like to express my sincere gratitude to Dr Erik de Ruyter van Steveninck for the Dutch translation of the summary of this dissertation.

I appreciate colleagues at UNESCO-IHE such as Jeremiah Kiptala, Dr Chol Deng Thon Abel, Veronica Minaya, Graciela Alvarez, Adeboye Omatayo, Amadou Keita, Dr George Lutterodt, Dr Fidelis Kilonzo and Dr Njenga Mburu with whom I shared frustrations, excitements, sought informal advice and relaxed with away from the rigors of dissertation work; they made the PhD experience to be social, enjoyable and fulfilling. I am grateful to the many graduate students whom I worked with in the Mara and the enriching discussions we had; Elvirah Riungu and Zipporah Gichana (University of Eldoret), Evance Mbao, Cheplabatt Kabon and Audrey Tsitsiche (Egerton University), and Jessica Salcedo Borda and Anne Lilande (UNESCO-IHE). I am also grateful for the dedicated and timely assistance of Ms Jolanda Boots at the UNESCO-IHE fellowship office.

I appreciate the funding sources that made this dissertation possible. Firstly, the Dutch Ministry of Foreign Affairs through UNESCO-IHE Partnership Research Fund (UPaRF) for funding the MaraFlows project through which I was offered this PhD fellowship. I would like to appreciate the USAID MSc scholarships through the GLOWS project that contributed to some MSc studies under my dissertation work, and the partial support provided by AFRIVAL (ERC-Stg 240002) for the stable isotopes analysis at KU Leuven. I would also like to acknowledge the Kenya National Council for Science and Technology for granting me and my supervisory team permission to conduct this research. I appreciate the TransMara Conservancy and the Narok County Council (Narok County) for granting me access into the Mara Triangle and Masai Mara National Reserve, respectively.

To my lovely wife Nyatichi, thank you so much for all your support and patience, and for reminding me that there was more to life than just this dissertation. Special and profound appreciation for taking care of our son Bartundo during the long days spent away from home. Special thanks to my father, brothers and sisters for their love and encouragement.

Delft, November 2014

Frank Onderi Masese

Summary

To properly conserve, restore and manage riverine ecosystems and the services they provide, it is pertinent to understand their functional dynamics. However, there is still a major knowledge gap concerning the functioning of tropical rivers in terms of energy sources supporting riverine fisheries. I reviewed the anthropogenic influences on organic matter processes, energy sources and attributes of riverine food webs in the Lake Victoria basin, but also expanded the review to incorporate recent research findings from the tropics. Contrasting findings have been presented on the diversity of shredders and their role in organic matter processing in tropical streams. Recent tropical research has also highlighted the importance of autochthonous carbon, even in small forested streams. However similar studies are very limited in African tropical streams making it difficult to determine their place in emerging patterns of carbon flow in the tropics.

This study was conducted in the Mara River, which is an important transboundary river with its headwaters in the Mau Forest Complex in Kenya and draining to Lake Victoria through Tanzania. In its headwaters, the basin is drained by two main tributaries, the Amala and Nyangores Rivers which merge in the middle reaches to form the Mara River mainstem. The overall objective of this dissertation was to better understand the functioning of the Mara River by assessing the spatio-temporal dynamics of organic matter sources and supply under different land-use and flow conditions and the influence of these dynamics on energy flow for consumers in the river. I collected benthic macroinvertebrates from open- and closed-canopy streams and classified them into functional feeding groups (FFGs) using gut content analysis. In total I identified 43 predators, 26 collectors, 19 scrapers and 19 shredders. Species richness was higher in closed-canopy forested streams where shredders were also the dominant group in terms of biomass. Seven shredder taxa occurred only in closed-canopy forested streams highlighting the importance of maintaining water and habitat quality, including the input of leaf litter of the right quality, in the studied streams. The findings suggest that Kenyan highland streams harbor a diverse shredder assemblage contrary to earlier findings that had identified a limited number of shredder taxa.

I subsequently used the composition of macroinvertebrate functional feeding groups (FFGs) and the ecosystem process of leaf breakdown as structural and functional indicators, respectively, of ecosystem health in upland Kenyan streams. Coarse- and fine-mesh litterbags were used to compare microbial (fine-mesh) with shredder + microbial (coarse-mesh) breakdown rates, and by extension, determine the role of shredders in litter processing of leaves of different tree species (native *Croton macrostachyus* and *Syzygium cordatum* and the exotic *Eucalyptus globulus*). Breakdown rates were generally higher in coarse- compared with fine-mesh litterbags for the native leaf species and the relative differences in breakdown rates among leaf species remained unaltered in both agriculture and forest streams. Shredders were relatively more important in forest compared with agriculture streams where microbial breakdown was more important. Moreover, shredder mediated leaf litter breakdown was dependent on leaf species, and was highest for *C. macrostachyus* and lowest for *E. globulus,* suggesting that replacement of indigenous riparian vegetation with poorer quality *Eucalyptus* species along streams has the potential to reduce nutrient cycling in streams.

To study organic matter dynamics is these streams, I assessed the influence of land use change on the composition and concentration of dissolved organic matter (DOM) and investigated its links with whole-stream ecosystem metabolism. Optical properties of DOM indicated notable shifts in composition along a land use gradient. Forest streams were associated with higher molecular weight and terrestrially derived DOM whereas agriculture streams were associated with autochthonously produced and low molecular weight DOM and photodegradation due to the open canopy. However, aromaticity was high at all sites irrespective of catchment land use. In agricultural areas high aromaticity likely originated from farmlands where soils are mobilized during tillage and carried into streams and rivers by runoff. Gross primary production (GPP) and ecosystem respiration (ER) were generally higher in agriculture streams, because of slightly open canopy and higher nutrient concentrations. The findings of this study are important because, in addition to reinforcing the role of tropical streams and rivers in the global carbon cycle, they highlight the consequences of land use change on ecosystem functioning in a region where land use activities are poised to intensify in response to human population growth.

Lastly, I used natural abundances of stable carbon ($\delta^{13}C$) and nitrogen ($\delta^{15}N$) isotopes to quantify spatial and temporal patterns of carbon flow in food webs in the longitudinal gradient of the Mara River. River reaches were selected that were under different levels of human and mammalian herbivore (livestock and wildlife) influences. Potential primary producers (terrestrial C3 and C4 producers and periphyton) and consumers (invertebrates and fish) were collected during the dry and wet seasons to represent a range of contrasting flow conditions. I used Stable Isotope Analysis in R (SIAR) Bayesian mixing model to partition terrestrial and autochthonous sources of organic carbon supporting consumer trophic groups. Overall periphyton dominated contributions to consumers during the dry season. During the wet season, however, the importance of terrestrially-derived carbon for consumers was higher with the importance of C3 producers declining with distance from the forested upper reaches as the importance of C4 producers increased in river reaches receiving livestock and hippo inputs. This study highlights the importance of large mammalian herbivores on the functioning of riverine ecosystems and the implications of their loss from savanna landscapes that currently harbour remnant populations.

The results of this dissertation contribute data to discussions on the effects of land use change on the functioning of upland streams and food webs in savanna rivers with regard to carbon flow and the vectoring role played by large mammalian herbivores as they transfer terrestrial organic matter and nutrients into streams and rivers. This study also provides information and recommendations that will guide future research and management actions for the sustainability of the Mara River and linked ecosystems in the Lake Victoria basin.

Contents

List of abbreviations and acronyms

AFDM	Ash-Free-Dry Mass
AGR	Streams in agriculture land use
ANCOVA	Analysis of covariance
ANOVA	Analysis of variance
APHA	American Public Health Association
BIX	Biological Autochthonous Index
CPOM	Coarse Particulate Organic Matter
DEM	Digital Elevation Model
DO	Dissolved Oxygen
DOC	Dissolved Organic Carbon
DOM	Dissolved Organic Matter
EAFRO	East African Fisheries Research Organisation
EDM	Energy Dissipation Model
EEM	Excitation Emission Matrix
ENSO	ElNino Southern Oscillation
ER	Ecosystem Respiration
FAO	Food and Agriculture Organisation
FBOM	Fine Benthic Organic Matter
FFG	Functional Feeding Group
FOR	Streams in forest land use
FPC	Flood Pulse Concept
GCA	Gut Contents Analysis
GLM	General Linear Model
GoK	Government of Kenya
GPP	Gross Primary Production
Ha	Hectares
HDPE	High Density Polyethene
ICP-MS	Inductively Coupled Plasma - Mass Spectrophotometer
ITCZ	InterTropical Convergence Zone
JICA	Japan International Cooperation Agency
LVB	Lake Victoria Basin
LVBC	Lake Victoria Basin Commission
MANOVA	Multivariate Analysis of Variance
MEA	Millennium Ecosystem Assessment
MFC	Mau Forest Complex
MIX	Streams in mixed land use
MMNR	Masai Mara National Reserve
NBI	Nile Basin Initiative
NEP	Net Ecosystem Production
NMDS	NonMetric Multidimensional Scaling
NTU	Nephelometric Turbidity Units
OM	Organic Matter

PP	Primary Production
P/R	Production/Respiration
PAST	Paleontological Statistics
PCA	Principal Component Analysis
POM	Particulate Organic Matter
RCC	River Continuum Concept
RDS	River Distance
RPM	Riverine Productivity Model
SIA	Stable Isotopes Analysis
SIAR	Stable Isotopes Analysis in R
SIMPER	Similarity Percentages
SLR	Simple Linear Regression
SNP	Serengeti National Park
SRP	Soluble Reactive Phosphorus
SUVA	Specific Ultraviolet Absorbance
TDN	Total Dissolved Nitrogen
TEF	Trophic Enrichment Factor
TN	Total Nitrogen
TP	Total Phosphorus
TrPos	Trophic Positions
TSS	Total Suspended Solids
WQBAR	Water Quality Baseline Assessment Report
WWF	World Wide Fund for Nature

Chapter

1

1.1 General introduction

The character of a catchment fundamentally influences abiotic and biotic patterns and processes in streams and rivers. Since this thesis was popularized by Hynes (1975) many studies have established intimate linkages between terrestrial and riverine ecosystems (Vannote et al., 1980; Junk et al., 1989; Baxter et al., 2005). However, developing landscapes to meet human needs has altered surface water hydrology, geomorphology and physico-chemistry, impacting the ecology of streams (Allan, 2004; Nilsson et al., 2005; Dudgeon et al., 2006; Vörösmarty et al., 2010). Human activities in catchments and along riparian corridors have replaced natural forests with agriculture, pastures and exotic forestry species (Ferreira et al., 2006; Hladyz et al., 2011). These changes typically reduce habitat complexity and biodiversity, and affect organic matter dynamics, nutrient cycling, water purification and erosional processes (Palmer and Filoso, 2009; Acuña et al., 2013).

Riverine ecosystems exhibit heterogeneity in environmental conditions at multiple temporal and spatial scales ranging from microhabitats to whole landscapes (Frissell et al., 1986; Poff et al., 1997). Tropical streams and rivers are highly dynamic driven by flow pulses that affect connectivity with lateral habitats (Lewis, 2008). Discharge variability also alter physical habitat, water temperature, and the composition of biological communities and ecosystem functioning (Tockner et al., 2000; Fisher et al., 2001). The quality and quantity of leaf litter inputs into these streams is also seasonally variable and dependent on catchment and riparian conditions (Wantzen et al., 2008).

Land use and land cover changes represent accelerating threats to aquatic ecosystems in many parts of the world (Lambin and Geist, 2006; Odada et al., 2009). As a corollary to this, freshwaters are among the most threatened habitat types in the world, illustrated by a 50% decline in the Living Planet Index for freshwater ecosystems (an indicator of trends in populations of vertebrate species) between 1970 and 2000 (MEA, 2005). As anthropogenic land disturbance continues to increase, it is imperative to understand the functional links between land cover and land use changes and the physicochemical conditions, biological communities and the functioning of streams and rivers (Sutherland et al., 2002).

In Kenya, land use changes on various catchments and water towers have been increasingly characterized by human settlement, deforestation and wetland reclamation (Mati et al., 2008;

Pellika et al., 2009; Mango et al., 2011). Moreover, deforestation and land-use changes have led to a decline in terrestrial biodiversity and the large herbivorous mammals have particularly been affected (Ogutu et al., 2011). Land use changes are likely to provide opportunities for many invasive plants along stream margins, and in many catchments and along riparian corridors of many streams exotic *Eucalyptus* spp. have been introduced (e.g., Magana, 2001; Pellika et al., 2009). On developing landscapes, elevated concentrations of nutrients and sediments have been recorded in streams draining agricultural catchments during the wet season due to run-off from unpaved roads, footpaths and farmlands (Kitaka et al., 2002; Okungu and Opango, 2005; Kilonzo et al., 2013). In rural catchments, in-stream human activities (water abstraction, bathing, washing and watering of livestock) are influenced by weather conditions, being more common during the dry season (Mathooko, 2001; Yillia et al., 2008). With the increasing human population (averaging 3% p.a), these problems will likely be exacerbated jeopardizing efforts towards environmental management, biodiversity conservation and sustainable social and economic development.

1.2 Energy sources in riverine ecosystems

Studies that address energy sources and flow in riverine food webs are important to identify specific habitats and energy sources that underpin riverine productivity and can lead to improved management and restoration of rivers (Thorp et al., 2006; Naiman et al., 2012). Different theories have been put forth to explain the sources of energy (carbon) in riverine ecosystems and their spatial and temporal variation. The river-continuum concept (RCC; Vannote et al., 1980; Minshall et al., 1985) and the flood-pulse concept (FPC; Junk et al., 1989) were among the earliest models to be conceptualized to help explain the functioning of riverine ecosystems. The RCC was developed from observations on stable, unperturbed forested streams in the temperate region. The concept states that forested river systems have a longitudinal structure that results from a gradient of physical forces that change periodically along the length of the river as the stream size increases (Vannote et al., 1980). The RCC assumes that the importance of different sources of energy (allochthonous inputs, autochthonous production and transport of organic material from upstream) in a stream varies along the continuum and the relative abundance of macroinvertebrate functional feeding groups track the changes that occur (Vannote et al., 1980). On the other hand, the FPC (Junk et al., 1989; Bayley, 1995) applies to large floodplain rivers, especially in the tropics, and predicts that organic matter from upstream areas has negligible effects on production at higher trophic levels compared with the effects of organic matter produced and consumed locally on the floodplain.

2

Despite the significant contribution of these models to understanding the functioning of riverine ecosystems, their application to rivers in different biomes and climates around the world has faced some challenges. Even though popular, with over 6400 citations in Google Scholar as of August 2014, a number of discrepancies have been identified with the predictions of RCC over the years. These include discontinuities arising from tributaries (Osborne and Wiley, 1992), the concept does not apply to streams that lack riparian vegetation e.g., grassland or prairie streams (Wiley et al., 1990), its predictions relate only to the main channels of rivers and omit backwaters, marshes and floodplain linkages despite their importance in streams and rivers (Junk et al., 1989; Thorp and Delong, 1994, 2002), its applicability to large rivers, particularly tropical rivers with extensive floodplains (e.g., Welcome et al., 1989) and rivers from arid regions (Davies et al., 1994) has been questioned, and the concept fails to recognize that lotic systems are hierarchically organized, both spatially and temporally (e.g., Frissell et al, 1986; Naiman et al., 1988). Moreover, there have been discussions on the extent to which the RCC is applicable to tropical streams that lack a diverse shredder guild (Irons et al., 1994; Boyero et al., 2009; Dudgeon et al., 2010). Similarly, Tockner et al. (2000) proposed some changes on the FPC by suggesting an extension of the concept to include an interaction between temperature and flow. They also emphasize the ecological importance of expansion-contraction events below bankful flooding, especially in 'trained' temperate rivers, suggesting that 'flow pulses' should be incorporated as part of the 'flood pulse' concept (Tockner et al., 2000).

Sustained research efforts to address the discrepancies in the models, particularly the RCC (Vannote et al., 1980), have yielded considerable insight into carbon cycling and food web dynamics in riverine ecosystems. One such discovery was the "riverine heterotrophy paradox" in which scientists were intrigued by the realization that animal biomass in food webs in mid-sized rivers (> 4th order) could be fuelled largely by autochthonous autotrophic production (Meyer and Edwards, 1990; Lewis et al., 2001). This observation was captured in the 'riverine productivity model' (RPM; Thorp and Delong, 1994, 2002), which hypothesize that the major source of organic matter assimilated by animals in large rivers is derived from autotrophic production in the channel and allochthonous carbon inputs from the riparian zone. Recognizing that most dissolved organic matter (DOM) and particulate organic matter (POM) within channels are recalcitrant, and thus difficult to assimilate (Junk et al., 1989), the RPM suggests that labile autochthonous organic matter and moderately labile, direct organic subsidies from the riparian zone, compensate for the abundant but recalcitrant organic matter from upstream leakage and floodplains (Thorp and Delong, 2002; Thorp et al., 2006). The model also notes that while some production occurs in the main channel, major sources of carbon are derived from primary production in permanently inundated slackwaters and during regular and aperiodic inundation of the floodplain (Thorp and Delong, 2002).

Evaluating the relevance of these models for different river systems in different biomes around the world has been made particularly possible through the use of stable isotope analysis (SIA), which has become a standard method for quantifying contributions from different carbon sources to riverine consumers (Fry and Sherr, 1989; France, 1997; Vander Zanden and Rasmussen, 1999). In combination with gut content analyses (GCA), SIA has helped identify trophic interactions and food web attributes of aquatic ecosystems, including seasonal patterns (Vander Zanden and Rasmussen, 1999; Fisher et al., 2001; Post, 2002). However, much of our understanding of energy flow in riverine food webs is based on North (e.g. Herwig et al., 2007; Zeug and Winemiller, 2008) and South (e.g. Lewis et al., 2001; Hoeinghaus et al., 2007; Jepsen and Winemiller, 2007) American rivers, in addition to contributions from a range of river systems in Australia (e.g., Douglas et al., 2005; Hunt et al., 2012; Jardine et al., 2012). In comparison, little is known about carbon flow in African tropical streams and rivers.

There are discussions on the extent to which temperate based studies are representative of tropical streams and rivers (Boyero et al., 2009; Dudgeon et al., 2010). In the emergent discussions, organic matter processing and the relative importance of autochthonous and allochthonous sources of carbon in streams differing in their land use and land cover characteristics, sizes and degree of human influence have gained particular attention (e.g., Lau et al., 2009; Boyero et al., 2009, Dudgeon et al., 2010; Boyero et al., 2011c; Jinggut et al., 2012).

While much of organic matter and nutrient fluxes into streams occurs through direct litterfall from riparian vegetation and hydrologic transport through surface and sub-surface flowpaths, the movement of large mammalian herbivores can actively transfer organic matter and nutrients into rivers (Naiman and Rodgers, 1997; Polis et al., 1997; Grey and Harper, 2002; Bond et al., 2012). Animals deliver resource subsidies in recipient systems either in form of carcasses, in which the remains decomposes gradually fuelling nutrient uptake in the process, and through nutrient elimination via excretion and egestion (Kitchell et al., 1979, Wipfli et al., 1998; Vanni, 2002). Animal-mediated resource subsidies can facilitate energetic linkages between systems, and this can occur against naturally established gradients (Polis et al., 1997, Vanni, 2002; Jacobs et al., 2007; Marcarelli et al., 2011). Across the African tropical forests and savannas, large populations of wildlife were once part of the landscapes but land use and land cover changes over the years have led to significant declines in large herbivore populations, which have been replaced to a large extent by livestock (Prins, 2000). It is likely that the large wildlife facilitated terrestrial-aquatic food web linkages via delivery of resource subsidies in terms of nutrients (urine) and organic matter (faeces) during watering in the streams and rivers (Polis et al., 1997; Jacobs et al., 2007) but data is largely limited. Large animal inputs can influence nutrient cycling, ecosystem productivity and food web structure of the recipient ecosystem (Leroux and Loreau, 2008; Schmitz et al., 2010; Marcarelli et al. 2011). However, many of the studies on the influence of large animal subsidies on riverine ecosystems have been done in North American systems (Naiman et al. 1986, Post et al. 1998, Naiman et al. 2009, Walters et al. 2009). In eastern Africa, and in the Mara-Serengeti ecoregion, there is still a high density of large herbivores which offer a unique opportunity to study the influence of wildlife resource subsidies on riverine ecosystem function.

1.3 The Mara River basin

1.3.1 Location, physiography and climate

The trans-boundary Mara River basin (Kenya/ Tanzania) has a surface area of 13,835 km^2 and lies between latitudes 0°21'S and 1°54'S and longitudes 33°42'E and 35°54'E in Kenya (65%) and Tanzania (35%). The study area covers the Kenyan portion of Mara River basin (Figure 1.1). The Mara River has its source in Enapuiyapui Swamp on the Eastern Mau Escarpment, Kenya. In its upper part the catchment is drained by two main tributaries, the Amala and Nygangores Rivers, which flow through the forested Mau Forest Complex, tea plantations, settlements and small-scale agriculture. The two rivers converge to form the Mara River in a region characterized by large-scale agriculture (Plate 1.1). The river then meanders through two internationally renowned conservation areas, the Masai Mara National Reserve (MMNR) in Kenya (1718 km^2,) and the Serengeti National Park (SNP) in Tanzania (1741 km^2). In these protected areas two other main tributaries, the Talek River and the Sand River, join the mainstem Mara River (Figure 1.1). In its lower reaches the Mara flows through savannah grasslands and small scale agriculture that characterize the Serengeti region in Tanzania before emptying into Lake Victoria through the expansive Mara Swamp.

The Mara River basin is surrounded by mountainous and hilly topography in the northern and eastern parts with the elevation ranging from 3070 m above sea level (a.s.l.) in Mau Escarpment to about 1100 m a.s.l. in the Mara wetland. Both temperature and rainfall vary with altitude. The climate of the area is influenced by the inter-tropical convergence zone. The average mean temperature on the upper highlands is about 18°C whereas on the lowlands the average mean temperature is 25°C, but normally it ranges between 20°C and 27°C depending on the month of the year (Gereta et al., 2002). Rainfall varies from a high of around 1600 mm/yr in the forested uplands to around 850 mm/yr at the lower reaches. In

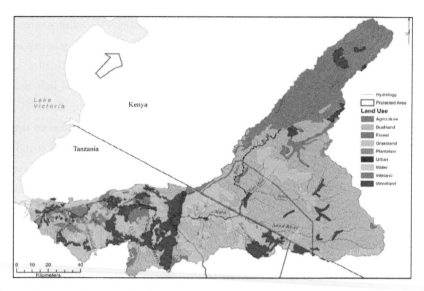

Figure 1.1. A map of the Mara river basin showing different land-use types, the main river and its main tributaries; Source, McCartney 2010.

addition to the spatial variability in rainfall, the catchment also experiences temporal variability with the different areas receiving variable amounts of rainfall over the year (Ogutu et al., 2007). The rainfall seasons are bi-modal, with the long rains expected from mid- March to June with a peak in April, while the short rains occur between September and December.

The geology of the area is predominantly made up of quaternary and tertiary volcanic deposits (Sombroek et al., 1982). The basin is dominated by two types of soils, the cambisols that occur in the middle and upper part of the basin, and the vertisols that are characteristic of the lower part (Sombroek et al., 1982). Cambisols have a number of characteristics that make them suitable for agriculture and dominate the upper catchments where broadleaf montane forests occur on the Mau Escarpment. These characteristics include good structural stability, high porosity, good water-holding capacity, good internal drainage, moderate-to-high natural fertility and an active soil fauna (Mati et al., 2008). Vertisols develop in expanding dark-colored clays commonly referred to as 'black cotton soil'. They are poorly drained because of the high clay content, and this makes them unsuitable for cultivation agriculture. These soil characteristics and the semi-arid conditions in the middle-reaches support savanna grasslands and bushland vegetation (Mati et al., 2008).

1.4 Problem statement and justification

Internationally, the Mara River is well known because of the annual migration of over 1 million wildebeests and other ungulates (zebra and gazelles) between MMNR in Kenya and SNP in Tanzania (Dobson et al., 2010). These conservation areas hosts one of the largest

6

Plate 1.1: Land use activities in the upper and middle reaches of the Mara River basin, Kenya. Upper reaches, (a) forest land use transitioning into (b) mixed small scale mixed agriculture (mainly grazing, maize, tea and agroforestry); middle reaches, (c) large-scale agricultural irrigation in the middle-ground and small-scale mixed farming (maize and grazing of livestock) in the back-ground, and (d) the savanna grasslands and scrublands in the Masai Mara National Reserve.

remnant savanna wildlife communities in the whole of eastern Africa and together with the linked ecosystems are major tourist attractions for both Kenya and Tanzania. The Mara is the only perennial river flowing through the region making it an important dry season source of water for sustaining wildlife in MMNR and SNP as well as the much-needed discharge for human use and water balance for Lake Victoria (Gereta et al., 2002; Mati et al., 2008; WREM, 2008; McClain et al., 2014).

As is the case for many river catchments in the African tropics, the Mara River basin is facing encroachment on protected forests and other fragile ecosystems for settlement and cultivation (Serneels et al., 2001; Mati et al., 2008) and uncontrolled water abstractions (Hoffman, et al., 2011). Increasing population and poor land management practices on newly converted lands have resulted in soil erosion and high sediment loads into the Mara River, pollution due to unregulated wastewater discharges, poor sanitation facilities and excessive use of agro-chemicals (WQBAR, 2007; Derfersha and Melesse, 2012; Kilonzo, 2014). A potential change in the flow regime of the river as a result of climate variability and land-use change has also been simulated (Melesse et al., 2008; Mango et al., 2011; Kilonzo, 2014). These alterations to natural conditions and disturbance regimes and the spatial arrangement of terrestrial vegetation and soils have important long-term consequences for the river and associated ecosystems (Serneels et al., 2001; Gereta et al., 2002; Mati et al., 2008). The loss of large populations of wildlife (mainly mammalian herbivores) through conversion of their grazing lands to agriculture (Serneels et al., 2001; Ogutu et al., 2011), likely reduces the supply of terrestrial organic matter and nutrients to rivers. Similarly, changes in vegetation type are likely to affect the quality and quantity of organic matter in the form of leaf litter in streams and the river with potential consequences on trophic resources and energy flow, but data are very limited.

Because of the intimate linkage between physical and biological processes in streams and rivers, direct modifications of natural flow regimes or indirectly through changing land-use or land-cover are manifested in the many aspects of river systems, including the physical and ecological aspects (Bunn and Arthington, 2002; Shafroth et al., 2010). In this regard, spatio-temporal dynamics in the source (quality) and supply of organic matter/ carbon, especially when these are also influenced by human activities, represent an important component of understanding the functioning of streams and rivers (Gawne et al., 2007; Lau et al., 2009; Hadwen et al., 2010b). For the Mara River, the importance of the river for downstream ecosystems in terms of water quality and quantity (Gereta et al., 2002) bears more credence to a need to understand linkages that will help in the development of appropriate management strategies. There is also a need to determine ecological responses in the structure of aquatic communities and food web structure along gradients of human and animal influences in the basin. With that understanding it will be possible to distinguish between human influences and natural causes when designing management programmes for the river and its catchment. Effective management of the Mara River basin, and those that share similar characteristics, requires up-to-date data on the influence of various land-use activities, the changing wildlife and livestock densities and flow variation on organic matter quality and supply and the corresponding responses in trophic relationships.

1.4 Objectives and research questions

The overall objective of this dissertation is to describe and contribute to the scarce data on the functioning of African tropical streams and rivers in terms of the role played by shredders on organic matter processing, spatial variability and seasonal dynamics of particulate and dissolved organic matter, ecosystem metabolism and the relative importance of allochthonous and autochthonous sources of carbon. The specific objectives of the study are:

1. To determine the influence of land use and dry-wet conditions on the diversity and distribution of macroinvertebrate shredders and their role on the processing of particulate organic matter (leaf litter) in the Mara River basin.

8

2. To determine the influence of land use and dry-wet conditions on the (a) quality and (b) quantity of dissolved and particulate organic matter and (c) ecosystem metabolism in the Mara River basin.

3. To determine the relative importance of allochthonous and autochthonous sources of energy for consumers along the longitudinal gradient of the Mara River.

4. To determine the seasonality of allochthonous and autochthonous sources of energy for consumers in the Mara River.

In order to achieve the stated objectives, a number of research questions were formulated to guide the study:

- What are the occurrence, abundance and distribution of macroinvertebrate shredders in streams in the Mara River basin?
- What is the influence of land-use, seasonality (dry and wet conditions) and shredder distribution on the processing of leaf litter in the basin?
- What is the influence of land-use and seasonality on the quality and quantity of dissolved organic carbon (DOC) and ecosystem metabolism in the basin?
- What is the relative importance of allochthonous and autochthonous sources of energy/ carbon for consumers along the Mara River?
- What is the influence of wet-dry seasons on the relative importance of allochthonous and autochthonous sources of carbon for consumers along the Mara River?

1.6 Dissertation structure

This dissertation consists of seven chapters. **Chapter 1** presents the general introduction by highlighting the current gaps in our knowledge of functioning of tropical streams. Additionally, it presents the problem statement and significance, research objectives and research questions of the study. A description of the study area is presented and includes information about the location, geology, climatic data and land use of the upper and middle-reaches of the Mara River basin, Kenya that is the focus of this study. **Chapter 2** provides a literature review of the influence of anthropogenic activities on the functioning of riverine ecosystems with a focus on the Lake Victoria basin in which the Mara River basin is part. Discussed in the chapter is the position of tropical streams and rivers in the global literature in terms of their food web attributes in the context of anthropogenic influences. The main findings of this dissertation are presented in **Chapters 3-6**. Each of these chapters consists of separate sections: abstract, introduction, materials and methods, results, discussion and conclusions. **Chapter 3-4** looks at the influence of both catchment and riparian land uses on organic matter input and processing in streams. **Chapter 3** looks at the diversity of macroinvertebrate functional feeding groups (FFGs) in Kenyan upland streams with a focus on shredders and their distribution in forest and agriculture streams during both the dry and wet seasons. This is important because tropical stream macroinvertebrates have not been extensively classified into FFGs, a situation that hampers studies on organic matter processing and assessments of the influences of land use change on the overall ecosystem integrity of streams and rivers in the region. **Chapter 4** uses the diversity and distribution of shredders and rates of organic matter (leaf litter) processing as indicators of the effects land use change and riparian conditions on the ecological health of the streams. The primary objective of **Chapter 5** is to assess the influence of land use change on the composition and concentration of dissolved organic matter (DOM) and its links to ecosystem metabolism and carbon cycling in the streams. Optical properties (fluorescence- and absorbance-based) are used to characterize changes in the composition of DOM in streams across a land use gradient from forestry to agriculture during both the dry and wet seasons. In **Chapter 6,** the influence of land use and herbivore-mediated subsidies on trophic sources for consumers in the Mara River, Kenya and a quantification of spatial and temporal patterns of carbon flow are investigated using natural abundances of stable carbon ($\delta^{13}C$) and nitrogen ($\delta^{15}N$) isotopes. Finally, **Chapter 7** synthesizes the main findings and proposes recommendations for management and future research.

Chapter

2

Anthropogenic influences on the structure and functioning of riverine ecosystems in the Lake Victoria basin, Kenya

Publication based on this chapter:
Masese FO, McClain ME. 2012. Trophic resources and emergent food web attributes in rivers of the Lake Victoria Basin: a review with reference to anthropogenic influences. Ecohydrology 5:685–707.

Abstract

The ecology of Lake Victoria and rivers draining its 180,000 km^2 basin has changed over the past century in response to growing anthropogenic influences that have altered basal resources, trophic status and interactions, and river flow regimes. Impacts on the ecology of the lake are well known, but little attention has focused on the ecological status of rivers supplying ecological services to a majority of the basin's over 30 million inhabitants. In this paper I review existing research on the ecological status of streams and rivers in the Lake Victoria Basin (LVB) and evaluate how they fit into emerging models of riverine ecosystem function in the tropics. Studies to date indicate that allochthonous sources dominate inputs to food webs in forested headwaters and savanna mid-reaches of rivers in the LVB, although transfer pathways vary with position in the river. While riparian vegetation phenology and hydrologic runoff pathways control the spatial and temporal fluxes of energy inputs in headwater streams, animals play increasingly important roles in savanna mid-reaches. Ecological patterns and processes in LVB rivers generally fit emerging models for the tropics, but studies completed to-date do not find autotrophic energy sources to be important. The findings of this review highlight the importance of specific management actions needed to safeguard services offered by the river to people and wildlife in the basin.

1.0 Introduction

Over the past 50 years, Lake Victoria has undergone fundamental changes to its ecology, the most well-known of which was the introduction of Nile Perch (*Lates niloticus* L. 1758, Centropomidae) in the 1950s and subsequent extinction of hundreds of species of native cichlids (Ogutu-Ohwayo, 1990; Witte et al., 1992). Twenty years before the introduction of the Nile Perch, however, limnological evidence already indicated increased nutrient loading and the beginning of a progressive shift in the trophic status of the lake (Hecky, 1993; Verschuren et al., 2002). Increased nutrient loading is correlated with development of the 180,000 km^2 catchment area of Lake Victoria, which covers portions of Tanzania, Uganda, Kenya, Rwanda, and Burundi. Since the 1930s, the human population of the lake catchment has increased from around 3.5 million to more than 30 million people, and land use change, expansion and intensification of agriculture, urbanization, industrialization, and freshwater abstractions have all increased proportionally (Bootsma and Hecky, 1993; Scheren et al., 2000; Verschuren et al., 2002).

Pronounced changes appeared in the lake during late 1970s when periods of anoxia were reported in the hypolimnectic waters (Hecky et al., 1994; Verschuren et al., 2002, Hecky et al., 2010). The mobilization of nutrients from catchment areas, and increased wet and dry

12

deposition over the lake brought about an increase in phytoplankton biomass and production, particularly in inshore areas and river mouths, but also in offshore areas (Hecky et al., 2010; Sitoki et al., 2010). As a result, the once mesotrophic lake (Talling, 1966) experienced a taxonomic shift from the historically persistent diatom *Aulacosira* Thwaites (Baciralliophyceae) to a highly frequent occurrence of cyanobacteria (Ochumba and Kibaara, 1989; Hecky, 1993; Gophen et al., 1995) and dominance of biologically fixed nitrogen (Mugidde et al., 2003; Hecky et al., 2010). There is also evidence that the eutrophication-induced deoxygenation of hypolimnectic zone may have contributed to the collapse of indigenous fish stocks through elimination of suitable habitat for certain deep-water species (Hecky et al., 1994; Verschuren et al., 2002), and increased turbidity also affected haplochromine populations that relied on vision for feeding and reproduction (Goudswaard et al., 2008). The lake is now considered to be eutrophic, with fluctuating water levels, high phosphorus and sediment loads and recurrent invasions of water hyacinth *Eichhornia crassipes* Mart. (Pontederiaceae) and hippograss (*Vossia cuspidata* Roxb. Griff. (Poaceae) (Scheren et al., 2000; Kolding et al., 2008; Offula et al., 2010).

Given the level and scale of change that human activities have had, and continue to have, on river catchments in the LVB, an immediate concern is the degree to which the ecological functioning of streams and rivers has been affected. The research attention devoted to ecological changes in the lake over a 50 year period is understandable given the lake's iconic status as the source of the White Nile River, its historical high diversity of fish species and the social and economic importance of its fishery, but on a regional scale more people rely on the freshwater and environmental services provided by the river corridors feeding the lake than on the lake itself. Assessing ecological changes in the rivers is difficult, however, because of a lack of information about river ecology prior to the considerable human influences that now shape the basin. Early research into the lake's ecology also considered the lower portions of tributary rivers, so we know that large numbers of lake fish species made periodic runs into tributary rivers to spawn during the rainy season (Graham, 1929; Whitehead, 1959), but little historical information exists for upper reaches of these rivers. Unlike the lake, however, which has been completely altered as a system, some streams and river sections in headwater forests and lowland savannas have been conserved to a degree that trophic status and food web attributes can be investigated in the absence of significant human alterations.

Studies on organic matter processing and food web structure in the LVB are also pertinent to a larger analysis of tropical riverine ecosystems and emerging evidence that they might be functionally different from their temperate counterparts (e.g., Boulton et al., 2008; Wantzen et al., 2008; Boyero et al., 2009; Dudgeon et al., 2010; Boyero et al., 2011a,b,c). In the ongoing discussions, comparisons have been made between temperate and tropical streams and rivers

13

in terms of organic matter processing, food web attributes and the major sources of energy fuelling metazoan food webs. Findings suggest that some of the predictions in the models developed in the temperate region may not be applicable to tropical streams and rivers.

Many studies in tropical headwater streams are supporting the importance of autotrophic production as a major source of energy for aquatic consumers (March and Pringle, 2003; Mantel et al., 2004; Brito et al., 2006; Li and Dudgeon, 2008; Coat et al., 2009; Lau et al., 2009). In small forest streams where allochthonous sources of energy and carbon dominate (Wallace et al., 1997), differences in the diversity of shredders have been reported between temperate and tropical streams. Shredders belong to a detritovore guild in forest streams where they help in the breakdown of coarse organic matter into fine detritus making it available to collectors and filterers, and thus playing a fundamental role in organic matter decomposition and nutrient cycling (Cummins et al., 1973; Wallace and Webster, 1996). Low diversity of shredders has been reported in New Guinea (Yule, 1996), New Zealand (Winterbourn et al., 1981), Hong Kong (Li and Dudgeon, 2008), East Africa (Tumwesigye et al., 2000, Dobson et al., 2002; Masese et al., 2009b), and the Neotropics (Greathouse and Pringle, 2006). Recent findings from a global assessment of the distribution of the detritivorous guild and its role in leaf litter decomposition in streams have reinforced some of these findings (Boyero et al., 2011a, b, c). Tropical riverine food webs have also been described as having shorter food chains with a dominance of omnivory and macroconsumers (Douglas et al., 2005; Coat et al., 2009). However, even with these emerging shared characteristics, it is arguable that within the tropics there is great heterogeneity in the structure and functioning of riverine ecosystems (Boulton et al., 2008). For instance, while some streams and rivers display a highly variable flow regime characterized by intermittence, others are more stable (Poff et al., 2006; Conway et al., 2009; Kennard et al., 2010). That notwithstanding, lack of comparable studies across regions and systems has restricted generalizations about determinants of ecosystem structure and function and predicting the likely impacts of global and regional human disturbances, notably land use change, anthropogenic climate change and the associated variations in the natural flow regimes of streams and rivers.

In part, lack of understanding on how ecosystems in Africa are functioning is an impediment to their sustainable management. In most cases, limited time series data on environmental and ecological conditions in African streams and rivers leads to a reliance on studies done in the temperate zone and elsewhere in the tropics to interpret existing information and develop management guidelines. In the LVB, information from other geographic regions has been used to answer questions about the status of the lake's fisheries or to interpret impacts of human activities (e.g., Kolding et al., 2008). While riverine ecosystems in the basin share a number of climatic characteristics with tropical counterparts where much ecological work has been done (e.g., Hong Kong, tropical Australia and the Neotropics), some clear differences also exist. For instance, African streams and rivers in arid and semi-arid savanna climates tend to be less predictable and more prone to unexpected floods and droughts as a result of pronounced spatial and temporal rainfall variability (Conway, 2002; Conway et al., 2009; Kizza et al., 2009). By learning from what has been done in tropical streams elsewhere and integrating it with the information available, we can improve our understanding of the functioning of streams and rivers in the LVB.

Literature on the functioning of aquatic ecosystems in the LVB is growing. A number of studies have looked at different aspects of streams and rivers which can be interpreted to give an account on their functioning and the effects of anthropogenic disturbance. This review explores the functioning of riverine ecosystems in the LVB in terms of major sources of energy, food web attributes, and flow regime. The specific questions addressed include: 1) What are the major sources of energy to streams and rivers in the region? 2) What are the emergent characteristics in food webs in streams and rivers? 3) How do these characteristics fit into emerging trends and models for tropical systems in Hong Kong, the Neotropics, Brazil and tropical Australia where intensive and long-term studies have been done? 4) What is the influence of human activities on these processes? Though much of the information that has been used to answer these questions is from literature on the LVB, some studies have also been included from rivers, lakes and wetlands in the larger East African region. This is because of the shared hydroclimatic and runoff characteristics in the region. Use has also been made of literature on tropical streams and rivers in Asia, Australia, South and Central America, especially in areas where information from the LVB is not adequate and conclusive. Studies that have been of great significance include those on food webs and their attributes (e.g., March and Pringle, 2003; Mantel et al., 2004; Douglas et al., 2005; Brito et al., 2006; Lau et al., 2009; Coat et al., 2009; Dudgeon et al., 2010), organic matter processing (e.g., Li and Dudgeon, 2008; Wantzen et al., 2008) and seasonal hydrological influences on organic matter dynamics and trophic relationships (e.g., Junk et al., 1989; Winemiller, 1990; Winemiller and Kelso-Winemiller, 2003; Hadwen et al., 2010a; Pusey et al., 2010; Arthington and Belcombe, 2011).

2.0 The Lake Victoria Drainage Basin

2.1 Hydroclimatic conditions

The Lake Victoria Basin (Figure 2.1) is the source of the White Nile River. It occupies an area of about 251,000 km^2, of which 69,000 km^2 is the lake surface. The altitude of the lake surface is about 1,135 m a.s.l while the basin is made of a series of stepped plateaus with an

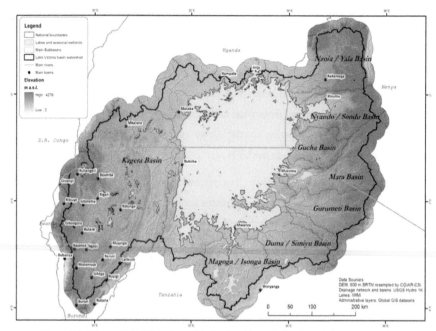

Figure 2.1. Map of Lake Victoria and its basin with an indication of the major rivers.

average elevation of 2,700 m a.s.l but rising to 4,000 m a.s.l or more in the mountainous areas. The lake has a mean depth of 40 m with a maximum depth of around 80–90 m.

The climate in the basin, ranging from wet tropical in the uplands to tropical savannah that characterizes much of the southern and eastern lowlands, is driven by the inter-tropical convergence zone (ITCZ), El Nino southern oscillation (ENSO), quasibiennial oscillation, large-scale monsoonal winds, meso-scale circulations and extra-tropical weather systems (Anyah and Semazzi, 2007). Over the lake, mean annual precipitation is around 1780 mm and mean annual evapotranspiration is around 1540 mm (Nicholson et al., 2000). In the LVB, a typical year has two peaks in rainfall, the long rains from March to May and the short rains from October to December (Rodhe and Virji, 1976; Kizza et al., 2009). The March-May rains contribute 39% of the total annual rainfall while the October-December rains contribute 26%, though there are considerable fluctuations (23%–50% for March-May rains and 15%–46% for October-December rains) (Kizza et al., 2009). This annual pattern of rainfall is also characteristic of the larger East Africa region (Rodhe and Virji, 1976). The rainfall varies from a minimum of around 800 mm in the lowlands and lakeshore areas to a maximum of around 2200 m in the highlands, with a catchment average of 1200 mm. In Kenya, the amount of rainfall varies spatially as a result of changes in relief features (Jaetzold and Schmidt, 1987). Mountainous areas, which are mostly humid and sub-humid, receive more rainfall while low-lying areas, including the areas adjacent to the Lake, receive less rainfall (Kizza et al., 2009), which is characteristic of the semi-arid tropical climate and savanna. However, some yearly and monthly variations have been reported over the past decade that are characterized by delays in the onset of rainfall and an increase in the duration and frequency of droughts (Ogutu et al., 2007). Extreme drought events occur approximately every 3–4 years during the short rains, every 7–8 years during the hot dry season (December to February), and every 5–8 years during the long rainy season (Awange et al., 2008). The main source of water into the lake is direct precipitation which accounts for approximately 80% of the water input (Sutcliffe and Petersen, 2007). River inflow and underground discharges contribute 20%, with the five major rivers, the Kagera, Nzoia, Gucha, Sondu-Miriu and Simiyu contributing over 60% (Table 2.1). About 75% of the water is lost through evapotranspiration, while about 18% flows out through the White Nile. The rest is lost through abstraction and groundwater losses. Thus, the inflow from the rivers is important for compensating the losses and maintaining lake levels.

 Riverine ecosystems in the LVB display highly seasonal hydrological regimes (e.g., Sutcliffe and Parks, 1999). Changes in some aspects of the natural flow regime of some rivers have also been reported (Conway, 2002; Tate et al., 2004). Some rivers have recorded an increase in the average daily discharge in the period 1990-2000 as compared with the period 1950-1960. The Rivers Sio, Nzoia and Yala recorded an increase in peak discharge of 8%, 31% and 38%, respectively (Sangale et al., 2005). While an increase in the average amount of rainfall between 1960 and 1990 of about 8% over much of the basin (Conway, 2002) may explain a portion of the recorded increases in peak discharge of the rivers, loss of native vegetation in the catchment areas has been identified as the most probable cause (Sangale et al., 2005). Increases in stream flows and reductions in interception and evapotranspiration losses have been measured in some catchments where native forest species have been replaced with pine plantations (JICA, 1980 cited in Sangale et al., 2005). Analysis of discharge records from catchments subjected to higher rates of deforestation have also been found to exhibit increased flood flows and decreased dry season baseflows relative to catchments with less deforestation (Melesse et al., 2008).

2.2 Aquatic biodiversity

The Lake Victoria and its tributary rivers are a major biodiversity hotspot. The lake itself supports more than 500 fish species, more than 90% of which are haplochromine cichlids Hilgendorf 1888 (Cichlidae) (Witte et al., 1992; Lowe-McConnell, 1987). However, more than 200 of the haplochromine cichlids are now considered to be extinct or threatened as a result of introductions of exotic fish species and ecological changes in the lake (Witte et al.,

Table 2.1. Catchment area sizes, mean human population density in the year 2000, rainfall and discharge (1950-2000) of major rivers in the Lake Victoria Basin; data from COWI 2002; Shepherd et al., 2000.

Country	River Basin	Catchment area (km²)	Human Population density/ km²	Mean annual rainfall (mm) [#]	Mean discharge (m³/s) [#]	Percentage (%) of total discharge[#]
Kenya	Nzoia	15143	221 (±154)	1492	116.7	14.8
	Nyando	3517	174 (±127)	1307	18.5	2.3
	Sondu-Miriu	3583	220 (±148)	1511	42.2	5.4
	Gucha	6612	224 (±183)	1519	58.0	7.5
	Sio	1450	-	1560	11.4	1.5
	Yala	3351	221 (±154)	1589	37.6	4.7
	South Awach	780	-	-	5.9	0.8
	North Awach	760	-	-	3.7	0.5
Kenya/ Tanzania	Mara	13915	46 (±56)	1040	37.5	4.7
Tanzania	Grumeti	13392	21 (±26)	879	11.5	1.4
	Mbalageti	3591	37(±22)	766	4.3	0.5
	E. Shore streams	6644	-	-	18.6	2.3
	Simiyu	11577	50(±26)	804	39.0	4.8
	Magoga-Muame	5207	88(±47)	842	8.4	1.0
	Nyashishi	1565	-	-	1.6	0.2
	Issanga	6812	48 (±22)	897	31.0	3.9
	S. Shore streams	8681	-	-	25.7	3.2
	Biharamulo	1928	-	-	17.8	2.2
	W. Shore streams	733	-	-	20.7	2.6
Burundi/Rwanda/ Uganda/ Tanzania	Kagera	59682	181(±196)	1051	260.9	33.5
Uganda	Bukora	8392	-	-	3.2	0.4
	Katonga	15244	-	-	5.1	0.7
	N. Shore streams	4288	-	-	1.5	0.2
Kenya, Tanzania, Uganda	Lake Edge	40682	133(±175)	-	1077	-

1992; Goudswaard et al., 2008). The basin is also rich in other taxa, including aquatic plants, aquatic-dependent reptiles, amphibians and mammals (Chapman et al., 2001; Aloo, 2003; Gichuki et al., 2001a; Munishi, 2007). The basin's waters support a diverse mollusc fauna comprising 54 species, with about one-fifth endemic to the basin (Brown, 1994). Macroconsumers in the basin include the atyid shrimp *Caridina nilotica* P. Roux 1833 (Atyidae) and freshwater crabs of the genus *Potamonautes* Macleay 1838 (Potamonautidae) (Goudswaard et al., 2006; Dobson, 2004). Aquatic obligate vertebrate species include five species of freshwater turtles, two aquatic snakes, monitor lizard (*Veranus niloticus* L., Veranidae), the Nile crocodile (*Crocodylus niloticus* Laurenti 1768, Crocodylidae), three species of otters, *Aonyx capensis* Schinz 1821, *A. congicus* Lönnberg 1910, and *Lutra maculicollis* Lichtenstein 1835 (Mustelidae) and hippopotamus or hippo (*Hippopotamus amphibius* L., 1758 (Hippopotamidae) (Oyugi, 2011).

Lake Victoria and its influent rivers historically supported a diverse and sustainable fishery (Graham, 1929; Whitehead, 1959). Prior to the insurgence of the introduced Nile Perch, cichlids formed more than 80% of the fish biomass with other endemic species forming the remaining fraction. Catches in the rivers were also diverse and seasonally augmented by periodic runs of anadromous fish species into rivers to breed and /or feed (Ochumba and Manyala, 1992; Manyala et al., 2005). The lower reaches and floodplains of large rivers and ecotonal zones of the lake supported large hippopotamus and crocodile populations. However, with unsustainable exploitation and conversion of their habitat the abundance of hippopotamuses and crocodiles started to decline, reaching a level of concern by early 1950s (EAFRO, 1955). Currently, crocodiles have become rare in the main lake and are confined to the lower reaches of the influent rivers.

In the last century there have been marked and at times rapid changes in the catchment and ecology of Lake Victorira. While changes in the ecology and biodiversity of the lake have been well documented, changes in the rivers have not been considered in such detail. The following review compiles what is known about the ecology and functioning of rivers of the LVB under different intensities of human influence.

3.0 Trophic resources and emergent food web attributes

Within the global context of energy flow and food web structure in riverine ecosystems, there is growing debate on the position of tropical streams and rivers (Boyero et al., 2009; Dudgeon et al., 2010). From long-term studies in the tropics, a synthesis has been presented on key attributes of streams and rivers in terms of their structure and function (Dudgeon, 2008). These attributes relate to the altitudinal distribution of aquatic biota, instream primary production, input of organic matter (leaf litter) and its processing. Studies that have examined energy flow and food web attributes have reported emergent characteristics that include a high degree of autochthony, widespread omnivory, short food chains and dominance of macroconsumers (Mantel et al., 2004; Douglas et al., 2005; Coat et al., 2009; Dudgeon et al., 2010). However, in the LVB, and East Africa in general, riverine ecosystems have not been examined to determine how they fit into these emerging trends.

3.1 Basal energy sources

Research to-date suggests that terrestrially-derived plant material is the predominant basal energy source driving in-stream metabolism in streams of the LVB (Omengo, 2010; Borda, 2012). Inputs of allochthonous material vary depending on the dominant vegetation type, season, hydrologic conditions, and position in the river network. In the LVB and its environs, a majority of river flow originates in streams arising from humid forested highlands with steep gradients, exceeding 60° in some areas. Because of the canopy cover, more than 90% in some areas, first and second order forested streams are cool with water temperatures rarely exceeding 16 °C (Omengo, 2010). Concentrations of dissolved organic matter average 3 mg/L (Omengo, 2010). The riparian vegetation is diverse (Mathooko and Kariuki, 2000; Dobson et al., 2002), but in some areas dominance in leaf litter input by one or two species is common. Retention of coarse particulate organic matter (CPOM) is high in forested streams (Mathooko et al., 2001; Morara et al., 2003; M'Erimba et al., 2006). Standing stocks have been reported at 280 gm^{-2} dry weight or higher (Dobson et al., 2002; Magana and Bretshko, 2003). The high CPOM standing stocks and diversity of plant species suggest a consistent input of basal energy of the right quantity for decomposers. However, spatial and temporal variations in standing stocks occur. Factors that contribute to spatial variation are mainly dependent on the type of riparian vegetation and the composition of retention structures that include debris dams, rock outcrops, roots, branches and stream banks (Mathooko et al., 2001; Morara et al., 2003). Temporal variations are influenced by the phenology of plants and discharge regime and season. Peaks in litter input occur during rain storms and during the dry season when some tree species shed all their leaves (Magana, 2001; Pers obser.). In savanna environments characteristic of the lower reaches of rivers in the LVB, litter inputs may be significantly reduced and animals may play an important role in the lateral transport of organic matter into streams and rivers (e.g., Jacobs et al., 2007).

The few studies that have examined the functional diversity and community structure of macroinvertebrate assemblages have concluded that while macroinvertebrates are diverse and abundant in streams of the region, shredders are limited (Dobson et al., 2002; Masese et al., 2009b). However, these studies have also indicated that identification of shredders is hampered by a lack of identification keys and the few studies that have been done have used keys (e.g., Merritt and Cummins, 1996) developed for temperate stream fauna. These keys have been found to be misleading for some taxa in the tropics (e.g., Dobson et al., 2002; Cheshire et al., 2005). While the limited number of studies that have been conducted to study shredders and their role in leaf litter processing in streams in the region might not be conclusive, the reported low number of shredder taxa has led to suggestions that microbial action and physical abrasion are more important (Mathooko et al., 2000; Dobson et al., 2003). The fraction of microbial carbon that makes its way to upper trophic levels is unknown for rivers in the LVB but assumed to be small based on available isotopic evidence from elsewhere (e.g., Bunn et al., 2003). This implies that other pathways of energy transfer from basal resources to higher trophic levels play major roles in energy flow among food webs. In the LVB, herbivorous fishes including *Haplochromis* Hilgendorf, 1888 (Cichlidae) spp., *Oreochromis variabilis* Boulenger 1906 and *O. leucostictus* Trewavas 1933 (Cichlidae)

(Ochumba and Manyala, 1992; Raburu and Masese, 2012) obtain a large proportion of their energy directly from plant materials. In so doing, they help in the breakdowm of CPOM into fine particulate organic matter (FPOM) that is consumed by invertebrates. As important prey for large piscivores such as *Bagrus docmak* Forsskäl 1775 (Bagridae) and *Schilbe intermedius* Ruppell 1832 (Schilbidae), these fishes also contribute directly to energy transfer to higher trophic levels.

On a longitudinal gradient, patterns in CPOM, FPOM and algal (both benthic and water column) standing stocks, vary with changes in stream order (Figure 2.2), although expected patterns may be obscured by human activities that have modified physical conditions in streams and rivers. For instance, as streams become bigger in size and canopy cover is reduced, primary production is expected to increase. This is recorded in increasing chlorophyll *a* concentrations in some rivers (Figure 2.2), but in others an increase in sediment load limits primary production and standing stock. The removal of riparian vegetation along low-order streams has also had the effect of increasing primary production by both macrophytes and phytoplankton while at the same time reducing inputs of CPOM. However, streams arising from and draining swampy areas where there is high diversity and abundance of macrophytes often record high levels of dissolved organic carbon, up to 15 mg/l relative to other streams (McCartney, 2010). Paradoxically, the same streams have recorded lower levels of nutrients; < 0.2 mg/l total phosphorus, < 1.5 mg/l total nitrogen and sometimes below detection levels of soluble reactive phosphorus (McCartney, 2010). This suggests a tight internal cycling of elements within these systems (Gichuki et al., 2001b; Machiwa, 2010).

The role of dissolved organic matter (DOM) in stream and river food webs in the LVB has not been investigated. DOM plays an important role in transferring carbon into the microbial food webs and subsequently into metazoan food webs (Findlay, 2003). However, this is dependent on the quantity and quality of DOM available. Labile DOM leached from fresh leaves is quickly metabolized by bacteria while recalcitrant DOM, mostly originating from adjacent soils is transported downstream. Even though the microbial loop is thought to be inefficient in terms of energy flow to higher trophic levels, its importance in tropical streams and rivers has gained more credence because of microbial breakdown of leaves, which is thought to be more important than that by shredders whose diversity is limited in the tropics (Irons et al., 1994). DOM is also a major source of carbon for biofilm growth in streams and rivers. Biofilms form the base of food webs supporting grazers such as crustaceans, insects, molluscs and some fish (Stevenson 1996; Douglas et al., 2005; Lefrançois et al., 2011).

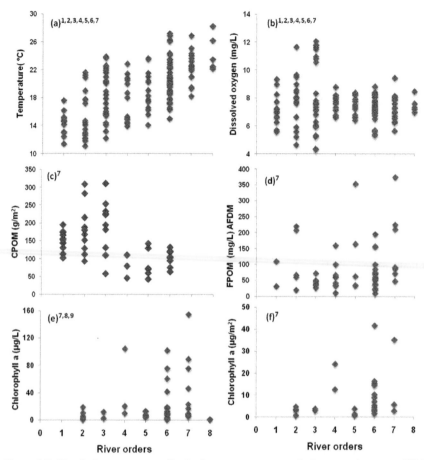

Figure 2.2. Trends in temperature, dissolved oxygen, coarse particulate organic matter (CPOM), AFDM fine particulate organic matter (FPOM) and chlorophyll *a* (both water column and benthic standing stocks) with stream order in the Lake Victoria basin, Kenya. Superscripts on figure numbers represent source of data; 1 = Masese et al., 2009a,b; 2 = Raburu et al., 2009; 3 = Kibichii et al., 2007; 4 = Raburu, 2003; 5 = Raburu PO, 2011 unpublished data; 6= Ojwang, 2006; 7= Masese FO, 2011 unpublished data, 8= NBI, 2009; 9 = Ojunga et al., 2010.

Studies examining downstream and lateral transport of organic matter in the LVB are limited. Studies in the Njoro River indicate the river exports approximately 641 kg of leaf litter to Lake Nakuru annually (Mathooko et al., 2001). Many streams in the LVB originate from or flow through macrophyte dominated swamps (Njuguna, 1996; Bavor and Waters, 2008; McCartney, 2010). This makes them important sources of dissolved organic matter for large rivers.

Transfers of energy between terrestrial and aquatic ecosystems and between lentic and lotic ecosystems in LVB are also mediated by animals (Jacobs et al., 2007; Strauch 2013). The role of large animals in savanna ecosystems is particularly noteworthy because among tropical systems, it is most pronounced in Africa. The Mara River, in particular, flows through a conserved landscape that supports millions of large herbivores that use the river system for watering. Because many Mara tributaries are seasonal and most of the water sources unreliable, ephemeral and dependent on local rainfall events, animals concentrate during the dry months around the perennial Mara River and the pools occurring in the bigger seasonal rivers. This in itself creates an opportunity for the animals to deposit large quantities of organic matter and high amounts of nutrients in the rivers (e.g., Hoberg et al., 2002; Jacobs et al., 2007; Strauch, 2013). During the rainy season, organic matter and nutrients deposited on the catchment and riparian areas are washed into the river system. In addition, there are more than 4000 hippopotamuses in the Masai Mara National Reserve stretch of the Mara River, representing an overall density of 27 hippopotamuses per km of river (Kanga et al., 2011). Hippopotamus transport large amounts of organic matter from riparian areas into large rivers daily. It is estimated than one hippopotamus can contribute about 9 metric tons of dry mass to aquatic bodies annually (Naiman and Rogers, 1997), although recent estimates are much lower than that (Subalusky et al., 2014). The activities of hippopotamuses also create many pools and lagoons in rivers and floodplains, which provide refuges for fish during the dry season (Naiman and Rogers, 1997). The pools and lagoons are subsequently fertilized by hippopotamus dung, which promotes primary and fish production (Hoberg et al., 2002; Mosepele et al., 2009). In the Mara River, the action of hippos stirring the water has been found to prevent formation of anoxic conditions when water levels decline in the dry season (Gereta and Wolanski, 1998; Wolanski and Gereta, 1999). These findings suggest that hippopotamuses and other wildlife contribute both directly and indirectly to the functioning of rivers in savanna landscapes of the LVB.

Fish migrations may also be important in transferring energy between lentic and lotic environments in the lower reaches of rivers in the LVB. Many fishes in Lake Victoria make periodic runs into influent rivers and their floodplains for breeding and/or feeding. The potamodromous fish species such as *Labeo victorianus* Boulenger 1901 and *Barbus altianalis* Boulenger 1900 (Cyprinidae), *Schilbe intermedius* (Schilbidae), *Synodontis victoriae* Boulenger, 1906 and *S. afrofischeri* Hilgendorf 1888 (Mochokidae), *Bagrus docmak* (Bagridae) and *Clarias gariepinus* Burchell, 1822 (Clariidae) migrate into tributary rivers in April-May and September-November (Lowe-McConnell, 1987; Ochumba and Manyala, 1992; Omondi and Ogari, 1994; Manyala et al., 2005), which are periods for the long and short rains, respectively. Some fish move up to 80 km upstream in the major rivers (Whitehead, 1959). Even though limited information exists on the contribution of potamondromous fishes in the supply of organic matter from Lake Victoria into its littoral areas and influent rivers, migrating fishes are known to be important allochthonous sources of organic matter and nutrients in other rivers (Flecker et al., 2010). The excretion of ammonia, other nitrogenous wastes and eggs that die are consumed during spawning runs and incorporated into heterotrophic pathways (Bilby et al., 1996; Flecker et al., 2010). Studies on anadromous salmonids have demonstrated the importance of migratory fishes in the recycling of elements between marine and freshwater habitats in North America (Kline et al., 1990; Cederholm et al., 1999).

3.2 Importance of allochthonous sources of organic matter in food webs

Food web studies in the LVB indicate that plant material is a primary source of basal energy in riverine ecosystems and suggest that allochthonous sources predominate. This is based on results from gut content analysis (GCA) and stable isotope analyses. The earliest food web studies were done in the lake itself (e.g., Worthington, 1933; Lêvêque, 1995). These studies indicate that the lake food-web is phytoplankton and detritus based. In tributary streams and rivers, GCA studies on various fish species revealed great diversity in the type of food items consumed. A study on non-cichlids by Corbet (1961) is considered to be the pioneering work on feeding habits of riverine fishes in the basin. This and later studies (e.g., Okedi, 971; Balirwa, 1979; Ochumba and Manyala, 1992) generally indicate that most species feed on plant material, including detritus, and some appreciable amount of insects. While the GCA methods used in these studies do not clearly distinguish between allochthonous and autochthonous carbon sources, they show that plant materials and detritus form the greatest bulk of food consumed by most fish species. In the Sondu-Miriu River, phytoplankton/algae were identified in the guts of only 4 of 18 species examined, while all species had large quantities of plant materials and detritus in their guts (Ochumba and Manyala, 1992). On comparing the food of *Barbus* spp. (Cyprinidae) from Lake Victoria and tributary rivers, Balirwa (1979) showed that riverine fish ingest more plant material than those in the lake, which is in agreement with earlier findings by Corbet (1961).

Carbon (δ^{13} C) and nitrogen (δ^{15} N; Tables 2.2 and 2.3) stable isotope data also indicate the predominance of terrestrial plant material in riverine food webs. The marked differences in the isotopic signature of primary food sources (phytoplankton, macrophytes, C4 and C3 plants) enables the identification of energy sources driving the food webs and their seasonal variations (Tieszen et al., 1979; Gichuki et al., 2001b; Ojwang' et al., 2007; Machiwa et al., 2010). In the LVB, stable isotope studies reveal that terrestrial plant material remains the predominant basal source of energy even as land use change alters the composition of riparian vegetation. A study investigating local watershed influences on energy sources for stream fishes found that when indigenous riparian vegetation was replaced with grasses (notably sugar cane), the C4 signature of the grass appeared in two fish species (*Labeo Victorianus* and

Barbus altianalis) investigated (Table 2.2 and 2.3; Ojwang' et al., 2007). Significant incorporation of C4 carbon sources into fish biomass in riverine ecosystems in the LVB is unexpected, as it is contrary to a number of previous studies showing very little organic carbon from C4 plants being incorporated into aquatic food webs (Bunn et al., 1997; Clapcott and Bunn, 2003). C4 carbon sources are considered to be of poor quality and contain high amounts of inhibitory compounds and chemicals (Clapcott and Bunn, 2003). It is therefore surprising how the C4 plants, especially sugar cane, have been able to contribute to stream food webs in the LVB. However, following the large amounts of detritus, mud and insects consumed by most fish species in the basin (Ochumba and Manyala, 1992) I support the suggestion by Ojwang et al. (2007) that the most likely pathway for carbon transfer from C4 plants to fish is through the detritus, biofilm and the macroinvertebrate food chains.

3.3 Importance of autochthonous sources of energy
Food web studies conducted to-date in streams and rivers of the LVB and adjacent Kenyan river basins have not identified autochthonous energy sources as important inputs, but research is limited and detailed isotopic studies of the kind conducted elsewhere still remain to be done. Studies in upland Kenyan streams indicate the presence of some primary production as evidenced by standing stocks of algae, as measured by chlorophyll *a* (Figure 2.2). However, much of this production is in open canopy agricultural streams as many forested streams are completely shaded and heteroptrophic (Omengo, 2010; Borda, 2012). In the LVB, there are indications that autochthonous production contributes to the food webs of streams and rivers, even though the exact amounts have not been determined. In the Sondu-Miriu River, quantitatively the most important food item consumed by fish is detritus (including mud) which contains some amount of benthic microalgae and cyanobacteria that were not identified in this study (Ochumba and Manyala, 1992). In one Rift Valley stream, diatoms were part of the detritus consumed by a freshwater crab of genus *Potamonautes* (Lancaster et al., 2008). Omnivores and species that consume large quantities of detritus have been found to derive most of their energy from algal components of the detritus (March and Pringle, 2003; Winemiller and Kelso-Winemiller, 2003; Brito et al., 2006) as a result of differential assimilation of the various components therein. Consequently, the contribution of autotrophic production to secondary production in the LVB could be higher than implied by Ochumba and Manyala (1992). Feeding on macrophytes and aquatic insects (some of them grazers/ scrapers) also increases the possibility of the fishes deriving higher energy values from autochthonous sources.

A growing number of studies in the tropics indicate that even in small forested streams autotrophic production is an important source of energy for aquatic consumers (Boyero et al.,

Table 2.2. Stable carbon isotope values of aquatic invertebrates and fish from a range of East African aquatic ecosystems compared with the isotopic values of aquatic algae and other primary sources. The number of samples (n) are shown in parentheses against the isotopic values. Superscripts on isotope values refers to the source of data.

Location	Primary sources of carbon	Algae	Primary and secondary consumers	Fish	Source
Upper Mara River	POM (agriculture land-use) -23.29±1.77 (23) POM (mixed land use) -27.53±1.03 (23) POM (forestry land use) -24.40 ±1.36 (23)				Omengo, 2010
Rivers Nzoia, Nyando, Sondu-Miriu and Yala	POC -21.17±3.02 (36) Debris -20.82±3.59 (13) Litter leaf -26.72 ± 2.30 (13) *Phragmites australis* -27.52 ±1.50(7) *Pistia stratiotes* -26.36 (1) *Eichhornia crassipes* -26.38±1.14 (6) C3 plants mean -24.34±3.64 (29) C4 plants mean -13.20±2.66 (33) Riparian vegetation (overall mean) -27.68 ±1.48 (16)	Algal mat -20.72 ±4.49 (5) *Ceratophylum* sp. -21.20 (1) Periphyton -22.63±2.93 (9)	*Caridina nilotica* -19.24 ±1.81 (16) Chironomid -15.64 Coleoptera -17.71 ±3.51 (5) Ephemeroptera -18.72 ± 1.57 (14) Crabs -16.83± 2.3 (11) Hemiptera -19.07 ±2.38 (11) Odonata -16.76 ±1.87 (6) Waterbug -23.89 ±2.57 (2)	**River Yala** *Bagrus docmak* -20.75 ±1.72 (10) *Barbus altianalis* -19.51±1.36 (30) *Labeo victorianus* -21.1±2.14 (43) **River Nzoia** *Bagrus docmak* -15.81±1.72 (15) *Barbus altianalis* -16.74±3.35 (61) *Lates niloticus* -17.65±0.77 (8) *Labeo victorianus* -15.42±3.19 (37) **River Sondu-Miriu** *Lates niloticus* -21.29±1.41 (14) *Barbus altianalis* -19.1±1.14 (20) *Labeo victorianus* -18.38 ±3.07 (27) **River Nyando** *Barbus altianalis* -15.5±4.76 (55) *Labeo victorianus* -15.02 ±4.21 (32)	Ojwang et al. (2007) Ojwang (2006)
[1]Mara River Mouth and Lake Victoria Nearshore	[1]Sediments -14.98±0.27[#] [1]Suspended particulate matter (SPM -24.53±1.55[#] *Cyperus papyrus* -12.7 to -11.9 (6)[#] -13.46 (1)[¥] *Phragmites* spp. -11.9 to -11.7 (4)[#] *Eichhornia crassipes* -27.8 to -27.2 (n=6)[#] -28.1 (2)[¥] *Typha* spp. -28.3 to -27.9 (4)[#] *Vossia* spp. -12.8 to -12.4 (2) -14.08 (1)[¥] Lake SPM -12.5 to -14.1 (20)[#]	Phytoplankton -19.9±0.7 (n=3) -24.1±0.6 (5)[¥] Filamentous algae -8.8 (1)[¥]	Zooplankton nearshore -17.6±0.8 (3) Zooplankton offshore -20.7 (4)[¥] *Caridina nilotica* -22.8 to -19.0 (23) -18.7 (9)[¥] Small mussels -24.1 to -23.4 (2) Prosobranch snail -18.9 (1) Ephemeroptera nymph -23.9±0.4 (10) -21.7 (2)[¥] Odonata nymph -27.0±1.1 (5) -26 (1)[¥] *Chaoborus* sp. -25.5 (2)[¥] Oligochaeta -23.8 (2)[¥]	*B. docmak* -21.3 (1) *Clarias gariepinus* -25.02 (1) *Haplochromis* spp. -24.1±0.5 (5) *L. niloticus* -22.2±1.2 (17) *Oreochromis niloticus* -22.3±1.9 (13) Immature *O. niloticus* -18.6±1.0 (4) *Protopterus aethiopicus* -23.8±0.5 (3) *Rastreonobola argentea* -23.8±0.5 (3) -20.0±1.2 (27) *Schilbe mystus* -27.4±1.5 (9) *Synodontis afrofischeri* -27.8±3.0 (7) *Tilapia Zillii* -19.2 (1) *Yssichromis laparogramma* -26.7 to -24.7 (2) -21.4±0.4 (38)[¥] *Y fusiformis* -20.8±0.7 (32)	[#]Machiwa 2010 Campbell et al., 2003 [¥]Ojwang et al., 2004

Table 2.3. Stable nitrogen isotope values of aquatic invertebrates and fish from a range of East African aquatic ecosystems compared with the isotopic values of aquatic algae and other primary sources. The number of samples (n) are shown in parentheses against the isotopic values. Superscripts on isotope values refers to the source of data.

Location	Terrestrial sources of carbon	Algae	Primary and secondary consumers	Fish	Source
Upper Mara River	POM (agriculture land-use) 6.08 ±1.31 (23) POM (mixed land use) 3.92 ±1.2 (23) POM (forestry land use) 4.71 ±1.35 (23)				Omengo (2010)
Rivers Nzoia, Nyando, Sondu-Miriu and Yala	POC 8.71± 1.38 (36) Debris 6.11±1.50 (13) Leaf litter 8.07 ±2.66 (13) *Phragmites australis* 5.49 ±1.45 (7) *Pistia stratiotes* 6.31 (1) *Eichhornia crassipes* 8.62 ±1.26 (6) C3 plants mean 7.06±1.9 (29) C4 plants mean 7.70±1.97 (33) Riparian vegetation (overall mean) 3.49±3.6 (16)	Algae mat 7.46±1.41 (5) *Ceratophylum* sp. 4.44 (1) Periphyton 7.38±1.99 (9)	*Caridina nilotica* 9.88 ±0.86 (16) Chironomid 8.33 Coleoptera 9.65±0.95 (5) Ephemeroptera 9.73 ±0.71 (14) Crabs 9.39±1.13 (11) Hemiptera 9.42 ±0.73 (11) Odonata 9.5 ±0.62 (6) Waterbug 8.44 ±1.95 (2)	**River Yala** *Bagrus docmak* 13.63±0.96 (10) *Barbus altianalis* 11.86±1.33 (30) *Labeo victorianus* 11.14±0.68 (43) **River Nzoia** *Bagrus docmak* 14.01±0.75 (15) *Barbus altianalis* 11.74±0.65 (61) *Lates niloticus* 8.95±1.22 (8) *Labeo victorianu* 10.67±0.88 (37) **River Sondu-Miriu** *Lates niloticus* 12.98±1.07 (14) *Barbus altianalis* 11.96±0.54) (20) *Labeo victorianus* 10.88 ±0.73 (27) **River Nyando** *Barbus altianalis* 11.36±0.96) (55) *Labeo victorianus* 10.77±0.98) (32)	Ojwang et al. (2007) Ojwang (2006)
[1]Mara River Mouth and Lake Victoria Nearshore	[1]Sediments -0.07±0.01[#] [1]SPM -0.56±0.02[#] *Cyperus papyrus* 0.2 to 2.7 (6)[#] 9.12 (1)[¥] *Phragmites* spp. 5.2 to 5.3 (4)[#] *Eichhornia crassipes* 2.7 to 3.4 (6)[#] 8.12 (1)[¥] *Typha* spp. 4.21±0.04 (4)[#] *Vossia* spp. 0.5 to 0.8 (2) 2.83 (1)[¥] Lake SPM 4.2 to -1.9 (20)[#] 4.7 to 10.1	Phytoplankton 4.3±0.8 (n=3) 4.1±0.3 (5)[¥] Filamentous algae 6.6 (1)[¥]	Zooplankton nearshore 7.1±0.3 (3) Zooplankton offshore 7.7 to 7.8 (2) 8.4±1.5 (4) *Caridina nilotica* 4.4 to 7.2 (23) 5.4±0.4 (9) Small mussels 7.1 to 7.8 (2) Prosobranch snail 5.6 (1) Ephemeroptera nymph 6.7±0.7 (10) 5.90 (2)[¥] Odonata nymph 6.0±0.4 (5) 11.20 (1)[¥] *Chaoborus* sp. 5.70 (2)[¥] Oligochaeta 3.90 (2)[¥]	*B. docmak* 12.7 (1) *Clarias gariepinus* 8.75 (1) *Haplochromis* spp. 9.2±1.1 (5) *L. niloticus* 11.8±1.4 (17) *Oreochromis niloticus* 9.8±1.1 (3) Immature *O. niloticus* 6.1±0.5 (4) *Protopterus aethiopicus* 10.6±0.4 (3) *Rastreonobola argentea* 11.3±1.4 (3) 12.1±0.6 (27) *Schilbe mystus* 9.5±0.7 (9) *Synodontis afrofischeri* 8.8±1.5 (7) *Tilapia Zillii* 8 (1) *Yssichromis laparogramma* 10.1 to 10.5 (2) 11.9±0.7 (38)[¥] *Y. yusiformis* 12.1±.6 (32)[¥]	[#]Machiwa 2010 Campbell et al., 2003 [¥]Ojwang et al., 2004

27

2009; Lau et al., 2009; Dudgeon et al., 2010). Autochthonous energy sources, especially periphytic microalgae and cyanobacteria, have been found to be very important to consumers in small shaded streams in the Neotropics (March and Pringle 2003; Coat et al., 2009), southeastern Brazil (Moulton et al. 2004, Brito et al. 2006), tropical Asia (Mantel et al. 2004; Li and Dudgeon, 2008; Lau et al., 2009; Dudgeon et al., 2010) and tropical Australia (Douglas et al., 2005; Hadwen et al., 2010a). In Hong Kong >60% of energy for some taxa is derived from periphyton and cyanobacteria (Lau et al., 2009; Dudgeon et al., 2010). This reliance on autochthonous production even in the presence of high allochthonous detrital inputs has been attributed to the high amount of inhibitory compounds in leaves of most tropical evergreen trees that makes them difficult to be broken down by terrestrial herbivores (Coley and Barone, 1996). The importance of autochthonous production in food webs of large tropical rivers has been attributed to filamentous and benthic microalgae, and to a lesser extent aquatic macrophytes (Pusey et al., 2010).

3.4 Occurrence and dominance of macroconsumers

Macroconsumers are defined as large-size consumers that are excluded by 2-mm mesh (*sensu* Pringle and Hamazaki, 1998). In low order streams in the LVB macroconsumers include fish (e.g., *Clarias* spp., *Barbus* spp.) and freshwater crabs (*Potamonautes* spp.). In the lower reaches of large rivers the atyid shrimp *Caridina nilotica* occurs, in addition to the crabs and fish. There are no indigenous species of crayfish on mainland Africa, but the introduced Louisiana red swamp crayfish *Procambarus clarkii* Girard 1852 (Cambaridae) has been reported in streams and wetlands in the upper catchment of rivers draining into the lake. The crayfish is established in Lake Naivasha where it was introduced in the 1970s and has had a number of ecological effects on the lake (Harper et al., 1995). However, studies that have examined the ecological role played by the atyid shrimp in streams and rivers in the region are lacking. In Kenyan streams *Potamonautes* spp. are widespread with high density (up to 25 individuals/m^2) and biomass (up to 4.6 g/m^2 dry mass in surber samples), often comprising >70% of the total macroinvertebrate biomass (Dobson et al., 2002; Dobson et al., 2007a,b). Most of these macroconsumers are omnivorous, which potentially increases leaf processing, either in concert with smaller macroinvertebrates or as a substitute for smaller macroinvertebrates. The dominance of *Potamonautes subikia* Cumberlidge and Dobson, 2008 (Potamonautidae) in a Rift Valley stream in Kenya and the low abundance of macroinvertebrates in the same stream have been identified as a possible top-down control exerted by the crab on stream macroinvertebrate communities (Cumberlidge and Dobson, 2008; Lancaster et al., 2008). Elsewhere in the tropics, a number of studies have reported a dominance of macroconsumers in stream food webs in the Neotropics (Pringle and Hamazaki,

1998; Coat et al., 2009), Brazil (Moulton et al., 2010) and tropical Australia (Douglas et al., 2005), but not in Hong Kong streams (Dudgeon et al., 2010).

3.5 Widespread omnivory

Because of the wide diversity of food items consumed by tropical fishes, food webs in rivers tend to be short, diffuse and with high connectance, rather than long and linear chains (Winemiller, 2004). In the LVB, studies in big rivers and Lake Victoria indicate that most species are omnivores that feed at more than one trophic level (Tables 2.2 and 2.3). Trophic enrichment of stable nitrogen isotopes is less than the predicted 3-4‰ (Minagawa and Wada, 1984). Even the earliest studies indicated that almost all fish species fed, to some extent, on insects, in addition to feeding on other food items (Corbet, 1961). This has also been observed in later studies both in the lake (Mbabazi, 2004) and the tributary rivers (Ochumba and Manyala, 1992; Raburu and Masese, 2012). For instance, in the Sondu-Miriu River nearly 40% of fish species fed on three or more different food items (Ochumba and Manyala, 1992). The most important food items were detritus (including mud), insects (adults and larvae), macrophytes and phytoplankton/ algae, consumed by 54%, 54%, 50% and 14% of the 28 fish species, respectively (Ochumba and Manyala, 1992). This trend is also witnessed today (Table 2.4), although there are minor variations than can be explained by seasonal changes in water quality and habitat conditions.

The earlier records of omnivory in the basin notwithstanding, there appears to be a tendency toward greater omnivory displayed by many fish species over the last half century (Ojwang et al., 2010). Some fishes that were hitherto piscivorous, such as *B. docmak* and *S. intermedius*, now appear to display greater inclination towards feeding at more than one trophic level (Goudswaard and Witte, 1997). Three species in the Sondu-Miriu River, which were considered to be predominantly piscivorous, *Lates niloticus, Bagrus docmak* and *Clarias gariepinus,* also consume a significant amount of other food items, including insects and molluscs (Ochumba and Manyala, 1992). The variation in food items consumed is so wide that two of the three species fall into different trophic groups in a recent trophic classification of riverine fishes in the LVB, with only *B. docmak* retaining the piscivorous trophic group (Raburu and Masese, 2012). In the Lake, the three species currently of greatest economic importance, *Lates niloticus* (introduced), *O. niloticus* L., 1757 (Cichlidae, introduced) and *R. argentea* Pellegrin 1904 (Cyprinidae, native) display some degree of omnivory (Wanink, 1998; Njiru et al., 2004). Even though most of the earlier studies used GCA to document food items consumed and classify fish species into different trophic levels, similar findings have been confirmed by the use of stable isotope analysis (Campbell et al., 2003; Mbabazi, 2004; Ojwang et al., 2004; Ojwang et al., 2010).

29

If indeed omnivory has been increasing in aquatic food webs in the LVB, then it is of interest to identify its cause(s) in light of the persistent and myriad environmental challenges in the lake and its basin (Verschuren et al., 2002). A number of factors determine the types of food items consumed and the extent to which resources are partitioned. Factors that determine the types of food consumed by fish include body morphology (Pusey et al., 1995), temporal and spatial changes in hydrologic conditions (Winemiller, 1990; Winemiller and Kelso-Winemiller, 2003; Winemiller and Jepsen, 1998; but see Pusey et al., 2010), habitat structure and food availability (Pusey et al., 1995; Rayner et al., 2009). Seasonal patterns of precipitation and hydrology influence the availability of habitats and food resources for tropical fishes (Lowe-McConnell, 1987; Winenemiller, 1989, 1990). In the LVB, river flow varies considerably within the year with flooding occurring during the peak of the rainy season, although year to year variations sometimes occur. In the Sondu-Miriu River, guts of most fish species were found to be empty during dry season low flows in March (Ochumba and Manyala, 1992). Though this was interpreted to signal the beginning of the breeding season when feeding in most fishes stop (Bishai and Abu-Gideiri (1965), it is likely that changes in the availability of preferred food resources, given the increased levels of degradation being experienced in the river by this time, might have contributed to this. In such a seasonally variable and degraded environment the most apt ecological strategy is one of omnivory by taking advantage of whatever food is available. For instance below industrial discharge points and in river sections that are severely degraded, omnivores are consistently more prevalent than other trophic groups, with some fish species displaying different feeding preferences depending on the environmental condition of the site inhabited (Raburu and Masese, 2012). It has also been observed that even *Labeo victorianus*, which is morphologically endowed with a subterminal mouth well suited for grazing aufwuchs from rocks in clearwater, has shifted to occasionally feed on insects and detritus (Ojwang, 2006), especially in river sections experiencing high turbidity levels (Pers. obser.). Consequently, opportunistic and omnivorous feeders (*Clarias* spp. and *Barbus* spp.) currently constitute the largest group in river systems in terms of diversity (e.g., Corbet et al., 1961; Ochumba and Manyala, 1992; Raburu, 2003; Raburu and Masese, 2012).

In the LVB omnivory has not been investigated in small streams. However, a study from elsewhere in the region indicated that omnivory is predominant for a freshwater crab of genus *Potamonautes* (Potamonautidae) which relies on a terrestrial subsidy of ants and detritus for its energy needs (Lancaster et al., 2008). This genus is widespread and abundant in LVB streams and rivers. Studies from Hong Kong indicate that omnivores are less common in small streams (Mantel et al., 2004; Dudgeon et al., 2010).

Table 2.4. Food items in the guts of different fish species along the Sondu-Miriu River during March, 2010.

Station	Fish species	*n*	Frequence of occurrence (%)[#]					
			Plant material	Detritus	Sand and mud	Algae	Insects	Mollusca
Changoi	*Barbus neumayeri* Fischer, 1884 (Cyprinidae)	6	33.3				100	
Kipranye	*B. neumayeri*	9	75		25		100	
	Clarias theodorae (Weber, 1897) (Clariidae)	3		100	100		100	66.6
	B. appleurogramma Boulenger, 1911	3	66.6		33.3		33.3	
Sondu	*B. altianalis* Boulenger, 1900	1		50	50		100	
	Pseudocrenilabrus multicolor Hilgendorf (Cichlidae)	2		50			50	
	Oreochromis variabilis Boulenger,1906	1	50				50	
	Tilapia zilli Garvais, 1848 (Cichlidae)	1		10	10		80	
Odino	*B. altianalis*	16	25	68.8		50	100	
	B. neumayeri	1					100	
	B. paludinosus (Peters, 1852)	3		100		100	33.3	
	B. apleurogramma	2		33.3			66.7	
	O. variabilis	9	22.2	100	11.1	66.7	11.1	
Wadh Langó	*B. altianalis*	20	85				50	10
	Labeo victorianus Boulenger, 1901 (Cyprinidae)	5		100	60			

[#]*Frequency of occurrence = total number of guts with food item/ total number of guts examined*100, n = number of fish in the sample. Where n = 1 the relative abundance of each food item is given. Data by Raburu PO (2010) unpublished.*

4.0 Human-induced changes in the Lake Victoria Basin

Prior to 1930, the human population in the LVB had been stable at around 3.5 million for over a century (Verschuren et al., 2002). However, the completion of the railroad from Mombasa to Kampala opened up the interior to settlement and mechanized farming of crops meant for the export markets. Because of immigration and improved health conditions, the population increased steadily with an annual growth rate of nearly 3% (Figure 2.3). Kenya has the largest population density in the basin with 38% of the total population. The average Kenyan population density in the basin is about 200 persons per km^2 compared with the national average of 38 persons per km^2. In some parts of the basin such as Kisii highlands and western Kenya, population density is more than 1000 persons per km^2. The rapidly increasing population has put pressure on the basin's resources, resulting in competition and conflicts over access rights and threats to sustainable use of resources (Odada et al., 2009). As catchments become more populated, deforestation and land use change have been among the many human activities that threaten water resources. All riparian countries, except Rwanda, have witnessed a decline in their forest cover over the last decade (FAO, 2010). In Kenya, it is estimated that only 1.7% of the country is forested, which is a drop from 10% in the 1950s (Figure 2.3). Much of the loss is associated with illegal logging, excisions and invasion of forests for subsistence farming, settlement, fuelwood harvesting and livestock grazing. In the upper Mara River Basin, 32% of the forest was lost between 1973 and 2000 alone (Mati et al., 2008). In the upper reaches of the Sondu-Miriu River Basin 21% and 10% of forests and bushland, respectively, were lost to farms and human settlement between 1986 and 2009 (Masese et al., 2011). Similar trends characterize other river basins in the region, including the Njoro, Nyando, Yala, Nzoia and Simiyu (e.g., Matiru, 2000; Raini, 2009; Twesigye et al., 2011). The same fate has befallen many wetlands along river valleys and floodplains and the lake's shores. Most have been encroached upon and drained for farming, settlement and grazing (Njuguna, 1996; Bavor and Waters, 2008; Muyodi et al., 2010; Twesigye et al., 2011).

Because of the high human population and its continued growth (Figure 2.3), waste disposal from municipalities has been a major source of pollution. Many small towns along the shores of the lake have no sewerage systems and those that have are operating beyond capacity. Industrial activities such as textiles, sugar manufacturing, alcohol distilleries, food processing, pulp and paper industry, and other informal sectors including brewing of the traditional beer 'Changaa', have been identified as sources of untreated wastewater into rivers and the lake (Bootsma and Hecky, 1993; Scheren et al., 2000; Verschuren et al., 2002; Nyenje et al., 2010). Mining activities in the catchments, including alluvial gold, have escalated. Storm water run-off from towns, agricultural areas and unpaved roads has often caused flooding, clogging of drainage channels and siltation of streams and rivers. These human activities have been identified as major concerns because of their potential to degrade water quality and interfere with the functioning of rivers, wetlands and the lake (Muyodi et al., 2010).

4.1 Effects on water quality

Recent studies have identified trends in the changes in Lake Victoria with human activities on the catchment (e.g., Verschuren et al., 2002; Hecky et al., 2010). Human population growth and loss of forest cover are related to deterioration in water quality both in rivers and in the lake (Figure 2.3). The current levels of nutrients and suspended sediments in most rivers are higher than when the human population was much less. For instance, in the Nzoia River, mean turbidity has increased from 40 NTUs in 1976 (Balirwa and Bugenyi, 1980) to the current level of 380 NTUs. In the Sondu-Miriu River, turbidity has more than doubled in 30 years from a mean of 60 NTUs in 1988 to the current mean of 130 NTUs. As a result, the rivers carry increased amounts of nutrients and sediment loads into the lake annually (Table 2.5). The water quality in rivers also deteriorates on a longitudinal

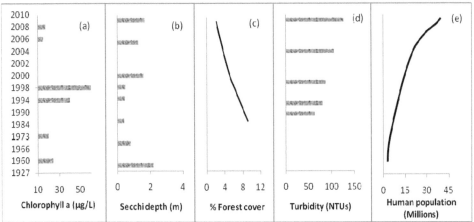

Figure 2.3. Historical trends in environmental conditions of the Lake Victoria and one of its influent rivers in relation to human-population growth and land use/ cover change. Figure letters present (a) and (b) are chlophyll *a* and secchi depth levels in the Lake Victoria, respectively, (c) forest cover in Kenya, (d) mean turbidity levels in the Sondu-Miriu River, and (e) human population growth in the Lake Victoria Basin. Sources of data (Graham, 1929; Talling, 1966; Akiyama et al., 1977; Ochumaba and Kibaara, 1989; Ochumba and Manyala, 1992; Muggide, 1993; Lehman and Branstrator, 1994; Lung'ayia et al., 2001; Silsibe et al., 2006; Mwashote and Shimbira, 1993; Raburu, 2003; Sitoki et al., 2010; Masese et al., 2011).

gradient with increases in turbidity, suspended sediments and nutrient levels (Figure 2.4). This reflects the distribution of human populations and activities on the Kenyan side of the basin. Human population is lower in the upper reaches because of protected forests and large-scale commercial farms. In downstream reaches, the population density increases together with intensification of agriculture on small plots. Many of the activities, for instance sugar cane processing and small-scale mining, are in the middle and lower reaches. However, despite these general trends, human activities are widespread and this has resulted in some low order streams in the upper reaches registering high nutrient, sediment and turbidity levels (Figure 2.3). Intensification of agriculture and the increased use of pesticides have led to contamination of water with pesticide residues (Getenga et al., 2004). Previous studies have reported high levels of heavy metals in water (Wandiga, 1981) and bottom sediments (Mwamburi, 2003) because of quarrying and alluvial mining of gold in the Mara, Yala and Gucha Rivers.

The water quality in the lake has also deteriorated over the years. Nitrogen loads from domestic sources have more than doubled. Nitrogen inputs from land run-off and rainfall have also increased (Scheren et al., 2000). While phosphorus input from land run-off marginally increased, that from domestic sources and rainfall increased by 300% and 6500%, respectively in 30 years (Scheren et al., 2000). The current input of phosphorus into the lake from rainfall and dry deposition accounts for

Table 2.5. Annual loads of total nitrogen (TN) and total phosphorus (TP) and sediment transport capacity index of major rivers in the Lake Victoria Basin. Values for TN and TP are averaged for the period 2001-2004. Data sources- COWI, 2002; Shepherd et al., 2000; NBI, 2005.

River Basin	TN (t/yr)	TP (t/yr)	Sediment transport capacity index	Average slope
Sio	248	47	-	-
Nzoia	3,340	946	0.14	2.3
Yala	999	102	0.14	2.3
Nyando	520	175	0.30	5.0
Sondu -Miriu	1,374	318	0.14	2.3
Gucha	2,849	283	0.16	2.0
Mara	1,701	304	0.15	2.0
North Awach	112	15	-	-
South Awach	322	39	-	-
Grumeti	561	185	0.12	1.6
Mbalageti	216	50	0.05	0.6
Simiyu	1,507	435	0.06	0.5
Magoga-Muame	278	50	0.05	0.5
Issanga	225	40	0.05	0.3
Kagera	29,303	1,892	0.24	3.0

nearly 80%, while biological fixation of nitrogen accounts for 78% (Muyodi et al., 2010). Several suggestions for high P deposition include soil exposure from deforestation, open field agriculture, and ash released from vegetation burning (Bootsma and Hecky, 1993). River inputs for total nitrogen and total phosphorus account for 4% and 17%, respectively (Muyodi et al., 2010). It has been estimated that 9.8 ktons (Mugidde et al. 2003) and 1.1 ktons of P inputs are from runoff and municipal wastewater discharges, respectively (Muggide et al., 2005). The input from sewerage wastewater, towns and cities around the lake is bound to increase with the rising human population in environments with limited waste handling and disposal facilities (Nyenje et al., 2010). Nutrient input into the lake has been linked to deoxygenation of hypolimnectic waters and diel super-saturation of surface water with oxygen during the day because of increased algal production (Hecky et al., 1994; Lung'ayia et al., 2001; Hecky et al., 2010; Sitoki et al., 2010). Cases of microbial contamination and cyanobacteria have also been reported with incidence of cholera and other waterborne diseases common among lake-water users (Muyodi et al., 2009). The transparency of the water has also decreased, partly because of algal blooms in the lake and silt carried in by the rivers (Lung'ayia et al., 2001; Sitoki et al., 2010). Records of orgonochlorides arising from agricultural activities in the catchments and near-shore areas have been observed (Wasswa et al., 2011). Heavy metals originating from alluvial gold mining and other geologic sources have been reported in lake sediments (Ongeri et al., 2009) and in fish tissues whereby long term consumption may pose a health risk to consumers (Oyoo-Okoth et al., 2011).

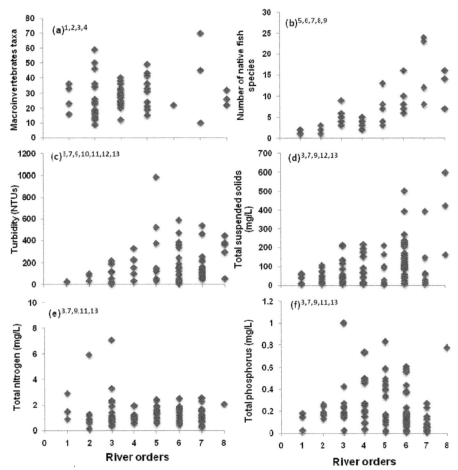

Figure 2.4. Longitudinal trends in physico-chemical variables and taxon richness of macroinvertebrates and fish in streams and rivers in the Lake Victoria basin and its environs during the last decade. Superscripts on figure numbers represent source of data; 1 = Masese et al., 2009a,b; 2 = Raburu et al., 2009; 3 = Kibichii et al., 2007; 4 = Maldonado et al., 2011; 5 = Ochumba and Manyala, 1992; 6 = Gichuki et al., 2001a; 7 = Raburu, 2003; 8 = Raburu and Masese, 2012; 9 = Raburu PO, 2011 unpublished data; 10 = Ojwang WO, 2006; 11 = McCartney, 2010, 12 = NBI 2009; 13 = Masese FO, 2011 unpublished data.

4.2 Ecosystem changes in rivers

Changes in land use and land cover in catchment areas have modified the composition and amount of material load in streams and rivers. These changes have significant implications for energy sources, flow and food web attributes both in streams and rivers (Figure 2.5). The current trend in the upper reaches of most streams and rivers has been that once the indigenous vegetation is cleared, riparian areas are planted with the exotic *Eucalyptus* L'Hér (Myrtaceae) species. Though other exotic species are also common, such as black wattle *Acacia mearnsii* De Wild (Fabaceae) and silky-oak *Gravillea robusta* A. Cunningham ex R. Br. (Proteaceae), *Eucalyptus* spp. are preferred by many farmers

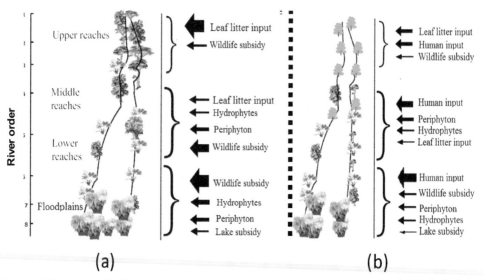

(a)   **(b)**

Figure 2.5. Human modifications on the longitudinal supply of organic matter into streams and rivers in the Lake Victoria basin. (a) represents various sources of organic matter prior to human changes on the catchments, and (b) represents changes in the supply of organic as a result of human activities. The size of the arrow represents the amount of organic matter supplied to the river (not actual amounts).

because of their fast growth and demand by the many tea factories that use them for heating their boilers. The landscape has changed with a dominance of *Eucalyptus* spp. along river valleys. Leaf litter from some exotics are of poorer quality (high C: N ratio) than the indigenous tree species they replace. Litter bag experiments in streams in the basin have indicated that *Eucalyptus globulus* Labill. (Myrtaceae) leaves, a common species across many catchments, decompose more slowly than native species (Chapter 4). Studies from elsewhere have highlighted the potential effects *Eucalyptus* spp. can have on water quality and the overall ecology of streams in the LVB. Their leachates have been found to increase conductivity and acidity, reduce oxygen content in water and inhibit reproduction in a freshwater fish (e.g., Canhoto and Laranjeira, 2007; Morrongiello et al., 2011). Because native trees have a different regime in the shedding of their leaves (Magana, 2001), their removal and replacement with exotic species has the potential to modify the amount and timing of leaf litter input to streams. The clearing of riparian vegetation along low order streams has also created favourable light conditions for the invasion of aquatic macrophytes in previously unvegetated channels (e.g., Julian et al., 2008). Growth of macrophytes has the potential to alter fluvial geomorphology, biogeochemical cycles and a number of biological processes (Wilcock et al., 2002; Julian et al., 2011).

The natural flow regime is important for maintaining ecological balance through regulating the composition and distribution of aquatic communities (Bunn and Arthington, 2002; Poff and Zimmerman, 2010). Some streams and rivers in LVB and its environs have recorded changes in their natural flow regimes (Shivoga et al., 2001; Melesse et al., 2008). For instance, freshets during the onset of rains help to flush streams and rivers by removing fines and, therefore, stabilize substrate for colonization by benthic organisms. Declines in river flows often lead to increases in water temperature and conductivity, decreases in oxygen levels and accumulation of fines. Abundance of taxa that are intolerant to these conditions, such as Ephemeroptera, Trichoptera and Plecoptera, reduce while the tolerant taxa, such as chironomids and oligochaetes, increase (Masese et al., 2009a,b; Mathooko et al., 2005). Extreme cases have been witnessed where stream reaches stop flowing, causing a decrease in diversity (Shivoga, 2001; Raini, 2009). In the lower reaches of the Sondu-Miriu River changes in the composition and abundance of phytoplankton communities have been observed (Mwashote and Shimbira, 1994).

Conversion of forest lands and human settlement on riparian areas, including the growth of towns, is changing the quality and flux of both particulate and dissolved fractions of organic matter. In the Mara River basin, there is a shift from high molecular weight to low molecular weight dissolved organic matter as a result of converting forests into agricultural land (Omengo, 2010). Increased primary production in streams in agricultural areas has been associated with reduced canopy cover and increased nutrient levels, especially dissolved nitrogen. In the Upper Mara River Basin, these changes have increased bioavailable organic matter and increased respiratory metabolism in streams across a land use gradient from forest reserve, through mixed catchments to agriculture (Omengo, 2010). On the longitudinal gradient, human-derived organic matter (domestic and industrial) dominates in place of upstream fluxes of dissolved and fine particulate fractions (Figure 2.5). In the middle and lower reaches of many rivers, the loss of wildlife has reduced a substantial subsidy of nutrients and organic matter, although livestock have compensated in some areas. Increased erosion and sediments have reduced periphyton growth and hydrophytes. In the floodplain wetlands at the mouth of rivers, the subsidy of nutrients and organic matter from the lake carried by potamodromous fishes in search of feeding and breeding areas has declined (Ochumba and Manyala, 1992; Manyala et al., 2005).

With streams and rivers transporting nutrients and sediments from catchment areas into the lake, their structure and function have been affected. The occurrence of phytoplankton genera such as *Nitzschia* Hass (Bacillariaceae), *Microcystis* Kützing, 1833 (Microcystaceae) and *Navicula* Bory (Naviculaceae) in the lower reaches of the Sondu-Miriu River more than 20 years ago (Mwashote and Shimbira, 1994) is an indication that the rivers have been experiencing high nutrient loads. These phytoplankton genera are considered to be indicative of nutrient pollution (Ojunga et al., 2010; Mpawenayo and Mathooko, 2005). In the lake, a shift from oligotrophy to mesotrophy was marked by the dominance of *Nitzschia* in the pelagic zone (Verschuren et al., 2002; Stager et al., 2009). Studies in the basin and its environs have documented changes in biotic attributes, which include reductions in abundance and richness of phytoplankton, macroinvertebrates and fish below municipal and industrial outfalls (Balirwa and Bugenyi, 1980; Raburu and Masese, 2012; Raburu et al., 2009; Ojunga et al., 2010).

A number of changes have also been observed among other communities in the basin's rivers. For instance, the once thriving riverine fishery (Whitehead, 1959; Balirwa and Bugenyi, 1980) has recorded major reductions in catches (Ochumba and Manyala, 1992; de Vos, 2000) and it is no longer sustainable. The hippopotamus and crocodile populations have also significantly plumeted, having lost their habitat to human activities. In this regard, the Mara River is an exception because of the protection offered by Masai Mara National Reserve and Serengeti National Park. On other catchments human-wildlife conflicts have been on the rise because of encroachment, with the hippopotamus particularly considered to be a 'nuisance' (Post, 2008; Kanga et al., 2012).

Modification of ecosystem linkages in the LVB by human activities is having negative impacts on ecosystem processes and biodiversity. In the lower reaches, construction of dikes along river banks has limited the interaction of rivers with their floodplains and the exchanges that occur (*sensu* Junk et al., 1989). Impoundment of the Sondu-Miriu River without providing a fish passage has stopped upstream migration of fish resulting in their congregation downstream of the dam (Raburu PO, 2011 unpublished data). Diversion of a significant fraction of river water for hydropower production deprives a long stretch of the Sondu-Miriu River of a significant amount of water. As a result, previously submerged habitats are exposed during base flow conditions, causing changes in macroinvertebrate assemblages (Raburu PO, 2011 unpublished data). Indications are that the once migratory fish species are adopting a sedentary mode of life in some sections of the rivers and the lake (Mugo and Tweddle, 1999; Ojwang et al., 2007). Recent surveys in the lake indicate that pollution sensitive species, such as the mormyrids, are showing particular associations with certain rivers, a sign that they are avoiding polluted rivers and areas within the lake (Ojuok, 2005). The loss of migratory fish populations is likely to negatively affect remnants of piscivorous animals (otters, monitor lizards and crocodiles) inhabiting the rivers, as well as people who depend on fish for a portion of their nutrition.

5.0 Management implications from the review

Environmental and ecological changes in the LVB are testing the resilience of riverine ecosystems and degrading environmental services enjoyed by people and wildlife of the region. A wide range of best management practices are required to preserve environmental services of rivers, and elimination of untreated waste discharges is particularly urgent, but this review also highlights three areas of action that should be emphasized in planning.

5.1 System connectivity

Longitudinal and lateral connectivity of rivers in the region is critical given the dependency of food webs on allochthonous sources of energy from upstream and adjoining riparian areas. Migrations of fish to and from Lake Victoria and along the river corridor are also important for accessing habitats and resources during specific phases of the life histories of many fish species. In the lower reaches of many rivers, it has been long recognized that the use of fishing gears that prevent the upstream migration of fish during the breeding season is detrimental to the fishery (Whitehead, 1959). This not only reduces the number of spawning populations but also potentially limits the organic matter and nutrient subsidy entering the river with migrating fish that then release nutrients when they defecate or die. Diking of rivers and draining wetlands by constructing channels has also limited the lateral connectivity and exchanges of energy and other materials between the rivers and their floodplains. Small and large-scale dam building in the LVB poses the greatest threat to longitudinal connectivity. A number of multipurpose dam-building projects are planned or proposed as part of regional development initiatives. These plans should be developed and implemented from a basin perspective that considers the optimal spatial configuration of dams and small-scale hydropower to minimize adverse impacts (Nel et al., 2009). Priority should also be given to off-channel storage of water for irrigation and other water supply purposes.

5.2 Environmental flows

Most countries of the LVB have water laws that recognize and protect environmental flows in rivers in addition to minimal flows required to meet basic human needs. In the 2002 Kenya Water Act, these protected flows are collectively referred to as the "Reserve", which is defined as "that quantity and quality of water required (a) to satisfy basic human needs for all people who are or may be supplied from the water resource; and (b) to protect aquatic ecosystems in order to secure ecologically sustainable development and use of the water resource." The fundamental role of hydrology in water quality and food web dynamics in the basin means that maintenance of aspects of the natural flow regime of streams and rivers (*sensu* Poff et al., 1997) is very important. To date, a comprehensive environmental flow assessment has been made in only one river of the basin, the Mara (LVBC and WWF-ESARPO, 2010). Recommended average monthly reserve flows from that assessment range from the 55th percentile of monthly flow duration curves during wet months of normal years and the 95th percentile of the annual flow duration curves during drought years (LVBC and WWF-ESARPO, 2010; McClain et al., 2014). By contrast, the default reserve flow generally applied by water managers irrespective of time of year is the 95th percentile (Q95). This is insufficient to meet the objectives of the law. Assessment of environmental flows for rivers in the region is, thus, critical in establishing the framework for managing the amount of water available and its allocation. In the Mara River, the environmental flow recommendations provide information that is necessary to sustain its ecosystem for the people and wildlife that depend on it (LVBC and WWF-ESARPO, 2010). Such studies are recommended for other rivers systems in the region, especially given the anticipated damming of many for hydropower production, water supply, and flood control.

5.3 Riparian corridors and remaining forests

River ecosystems in the LVB are dependent on energy inputs from riparian areas, and annual cycles in river ecological processes are coupled to those of riparian species. Considering the predominance of introduced plant species such as sugar cane, maize and *Eucalyptus* spp. on riparian areas, restoration and maintenance of indigenous riparian forests should be prioritized. Maintenance of buffer strips along river banks will also help to control sediments eroded from agricultural hillslopes. At the catchment level, deforestation should be checked and the current rehabilitation programmes expanded and strengthened. This will go a long way in controlling soil erosion and stabilizing river flows especially during the dry periods (Mango et al., 2011).

Conclusions

While additional research in LVB rivers has much to contribute to emerging models of tropical river ecology, the most pressing research needs in the region are, by far, those investigating the impacts of anthropogenic change on ecosystem functions and associated losses of ecosystem services used by people of the region. Priority here should be given to research focused on environmental flows and the impacts of reduced and altered flows on ecosystem function (Arthington et al., 2010; McClain et al., 2014). Ambitious regional plans for dam building, increased water abstractions for irrigation and domestic uses, and changes to runoff linked to land-use change all threaten to alter flow regimes in ways that likely impair ecosystem integrity. Research is needed into the vulnerabilities of species and processes to altered flows and the identification of environmental flow regimes that maintain a desired level of ecosystem functionality. A further priority is to investigate ecohydrological processes that support resource management objectives through natural assimilation of contaminants, attenuation of flood waves, or other processes that enhance the effectiveness of more engineered management systems (Zalewski, 2002).

Chapter

3

Macroinvertebrate functional feeding groups in Kenyan highland streams: implications for organic matter processing

Publication based on this chapter:
Masese FO, Kitaka N, Kipkemboi J, Gettel GM, Irvine K, McClain ME. 2014. Macroinvertebrate functional feeding groups in Kenyan highland streams: evidence for a diverse shredder guild. Freshwater Science 33: 435-450.

Abstract

Data on the functional composition of invertebrates in tropical streams are needed to develop models of ecosystem functioning and to assess anthropogenic effects on ecological condition. Macroinvertebrates were collected during dry and wet seasons from pools and riffles in 10 open- and 10 closed-canopy Kenyan highland streams. Collected macroinvertebrate were classified into functional feeding groups (FFGs), which were then used to assess effects of riparian condition and season on functional organization. Cluster analysis of gut contents was used to classify 86 taxa as collectors, predators, scrapers, or shredders while 23 more taxa whose guts were empty or had indistinguishable contents were classified according to literature. In total, 43 predators, 26 collectors, 19 scrapers, and 19 shredders were identified. Total abundance was higher in open-canopy agricultural streams, and species richness was higher in closed-canopy forested streams. Predators and shredders dominated richness and biomass, respectively, in the closed-canopy streams. The shredders, *Potamonautes* spp. (Brachyura: Potamonautidae) and *Tipula* spp. (Diptera: Tipulidae), made up >80% of total biomass in most samples containing both. Canopy cover and litter biomass strongly influenced shredder distribution. Seven shredder taxa occurred only in closed-canopy forested streams, and few shredder taxa occurred in areas of low litter input. Collectors dominated abundance at all sites. Richness and biomass of scrapers increased during the dry season, and more shredder taxa were collected during the rainy season. Temperate keys could not be used to assign some tropical invertebrates to FFGs, and examination of gut contents was needed to ascertain their FFGs. The Kenyan highland streams harbour a diverse shredder assemblage that plays an important role in organic matter processing and nutrient cycling.

3.1 Introduction

Macroinvertebrates are useful surrogates of ecosystem attributes, and the relative abundance of functional groups reflects anthropogenic impact (Merritt et al., 2002; Cummins et al., 2005, Merritt and Cummins, 2006). However, this approach is difficult to apply in many tropical streams where information on the functional composition of macroinvertebrate communities is limited (Boyero et al., 2009). Taxonomic keys developed for temperate-zone invertebrates (e.g., Merritt et al., 2008) often are used to assign tropical macroinvertebrates to trophic and functional feeding groups (FFGs). This approach has been successful for some taxa, but evidence is increasing that related species occurring in different regions do not share the same diets (Dobson et al., 2002; Cheshire et al., 2005; Chará-Serna et al., 2012). Even within regions, some taxa can shift their feeding in response to changes in land use and riparian conditions (Benstead and Pringle, 2004; Li and Dudgeon, 2008). Moreover, groups, such as the case-building Trichoptera, Plecoptera, and Gammaridae that dominate the detritivorous shredder guild in temperate streams are represented by very few taxa in tropical streams.

A number of authors have reported low diversity of shredders in streams in some tropical regions (e.g., Brazil: Gonçalves et al., 2006; Colombia: Mathuriau and Chauvet 2002; Costa Rica: Irons et al., 1994; East Africa: Tumwesigye et al., 2000, Dobson et al., 2002, Masese et al., 2009b; Hong Kong: Li and Dudgeon 2008; Papua New Guinea: Yule 1996), but others have reported diverse shredder assemblages (Australia: Cheshire et al., 2005; Malaysia: Yule et al., 2009; Salmah et al., 2013; Panama: Camacho et al., 2009). Resource availability and quality and biogeography may explain the paucity of shredders (Irons et al., 1994; Hallam and Read, 2006, Boulton et al., 2008), but growing evidence indicates that many tropical shredders have been overlooked because investigators commonly use temperate keys to assign FFGs (Dobson et al., 2002; Cheshire et al., 2005; Camacho et al., 2009). Scale and sampling effort also could be affecting taxon counts, as evidenced by the many shredder taxa (31) identified from 10 kick samples (sampling time for each ~2 min) from each of 52 forested streams in 9 catchments distributed over the Malaysian peninsula (Salmah et al., 2013).

In many parts of the world, landuse change, particularly loss of riparian vegetation, has resulted in loss of diversity and major shifts in the structural and functional organization of macroinvertebrates in streams (Allan, 2004; Benstead and Pringle, 2004; Jinggut et al., 2012). Loss of riparian forests increases stream temperatures through loss of shade (Baxter et al., 2005), reduces inputs of leaf litter, and affects the relative differences between wet and dry seasons (Wantzen et al., 2008). Shredder taxa are particularly vulnerable to riparian deforestation because it eliminates their main source of food. Many shredder species are adapted to cold water and may be close to their thermal maxima in the tropics. Thus, they may be especially susceptible to increases in water temperatures (Irons et al., 1994; Boyero et al., 2011a).

Knowledge of the functional composition of invertebrates in tropical streams is important to understand organic-matter processing, energy flow, trophic relationships, and the management actions needed to minimize the impairment of ecosystem functioning (Benstead and Pringle, 2004; Dudgeon, 2010; Boyero et al., 2011b, c; Ferreira et al., 2012). Research in eastern African streams has increased over the last two decades (Tumwesigye et al., 2000; Kasangaki et al., 2008; Minaya et al., 2013), but understanding of the functional composition of aquatic invertebrates and consequence on ecosystem structure and function is limited. Moreover, paucity of taxonomic information on most aquatic invertebrates is a major hindrance to ecological research. The only detailed information on FFGs in eastern African streams and in African tropical streams was published by Dobson et al. (2002), who collected samples only during the dry season and analyzed gut contents and mouthparts of 11 and 44%, respectively, of macroinvertebrate taxa collected. They classified the rest of the taxa based on temperate-zone keys. They concluded that shredders were scarce but noted that allocation of taxa to FFGs might have been incorrect because of their use of temperate-zone keys. I used gut contents to classify macroinvertebrates from 20 Kenyan highland streams to FFGs and tested their responses to riparian conditions, availability of leaf litter, and season. I used ratios of the various FFGs as indicators of ecosystem attributes and to assess the ecological health of the streams (Merritt et al., 2002; Merritt and Cummins, 2006). I hypothesize that: 1) analysis of gut contents of macroinvertebrates in these streams will help refine classification of FFGs, 2) functional groups are seasonally variable, 3) riparian conditions and availability of leaf litter play important roles in the distribution and abundance of shredders, and 4) the ratios of the various FFGs can be used as surrogates for ecosystem attributes to assess the ecological condition of the streams.

3.2 Materials and Methods

3.2.1 Study area

The study was conducted in mid-altitude (1900–2300 m a.s.l), 1st- to 3rd-order streams draining the western slopes of the Mau Escarpment, which forms part of the Kenyan Rift Valley. A total of 20 sites were selected for study (Figure 3.1). The streams form the headwaters of the Mara River, which drains the tropical moist-broadleaf Mau Forest Complex (MFC) that is a source of many rivers draining into Lakes Baringo, Nakuru, and Victoria. The Mara River flows to Lake Victoria.

The MFC has been fragmented and reduced in size because of excisions for human settlement, coniferous-forest plantations, and large-scale cultivation of tea (Lovett and Wasser, 1993). This activity also has resulted in a loss of riparian vegetation along streams and rivers. However, some intact forest blocks remain in protected forest reserves and national parks (Lovett and Wasser, 1993; Chapman and Chapman, 1996). At their lower edges, the forest blocks are protected by tea plantations that were established in the 1980s to buffer against encroachment. People living in the adjoining areas are mainly involved in semi-intensive smallholder agriculture, characterized by cash crops (mainly tea), food crops (mainly maize, beans, and potatoes), and animal husbandry. A clear transition with a shift in vegetation cover and tree species composition exists between the protected forests and inhabited and farmed areas (Plate 3.1) where exotic *Eucalyptus* species dominate riparian vegetation along the streams and rivers.

Climate of the area is relatively cool and seasonal because of the altitude. The area is characterized by distinct rainy seasons and low ambient temperature that falls <10°C during the cold months of January and February. Annual precipitation ranges from 1000 to 2000 mm with a bimodal regime. Dry conditions exist in January to March and wet conditions occur from March to July and October to November, periods with long and short rains, respectively.

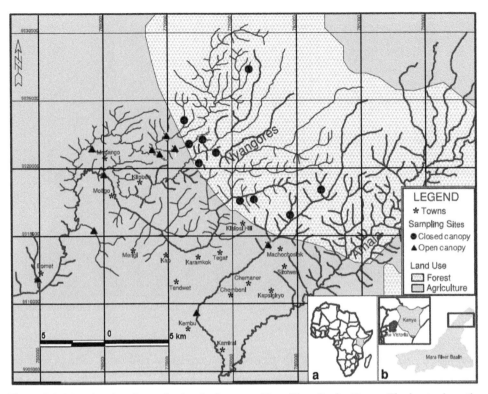

Figure 3.1. Map showing the study sites in the upper Mara River Basin, Kenya. The insets show the location of Kenya in Africa (a) and the location of the Mara River Basin in Kenya (b).

3.2.2 Field methods

Macroinvertebrates were sampled along 100-m reaches once during February 2011 (dry) and once during May to July 2011 (wet) at each site. Five benthic samples were collected at random locations in riffles and pools (total of 10 samples) with a dip net (300-μm mesh size), following Cheshire et al. (2005) and Yule et al. (2009), but with a shorter sampling time. An area covering ~30 × 50 cm was disturbed vigorously for 10 s, so as to avoid escape of large macroinvertebrates (Magana et al., 2012). All contents of the net were emptied into polythene bags in which they were preserved them using 75% ethanol, and transported to the laboratory for further processing.

At each site, % in-stream canopy cover, stream width, water depth, velocity, and discharge were measured over a 100-m reach. Stream width was measured with a measuring tape on 10 transects at midpoints of 10-m intervals along the reach. On each transect, water depth was measured with a 1-m ruler at a minimum of 5 points. Velocity was measured at the same points as depth with a mechanical flow meter (General Oceanics; 2030 Flowmeter, Miami, Florida). Stream discharge was estimated by the velocity–area method (Wetzel and Likens, 2000). At each point, the percentage of substratum covered by leaf litter (% leaf litter) was also estimated. The proportion of riffles and pools were

45

Plate 3.1: (a) Closed-canopy forest and (b) open-canopy agriculture streams in the upper Mara River basin, Kenya during the dry season

quantified as the proportion of the 10 transects that crossed a pool or a riffle. Concurrent measurements of pH, dissolved O_2 (DO), temperature, and electrical conductivity were made *in situ* with a YSI multiprobe water-quality meter (556 MPS; Yellow Springs Instruments, Yellow Springs, Ohio). A portable Hach turbidity meter was used to measure turbidity (2100P ISO Turbidimeter, Hach Company, Loveland, Colorado). All coarse particulate organic matter (CPOM) in dip-net samples was washed through a 100-µm-mesh sieve to remove macroinvertebrates and inorganic materials. The CPOM was dried to a constant mass at 68°C for ≥48 h, and the different fractions- leaves, sticks, seeds, and flowers- weighed separately to the nearest 0.1 mg with a Sartorius balance (SECURA224-1ORU; Sartorius, Goettingen, Germany).

3.2.3 Laboratory analyses

After sorting macroinvertebrates from debris, they were preserved in 75% ethanol and identified to the lowest-possible taxonomic level or morphospecies with the aid of keys in several guides (Day et al., 2002a,b,c; de Moor et al., 2003a, b; Stals and de Moor, 2007; Merritt et al., 2008). The wet mass of all macroinvertebrate individuals was used as an estimate of biomass. For most macroinvertebrates, diets were described from gut contents according to Cheshire et al. (2005). Gut contents of 3 (rare species) to 61 individuals were analyzed per species. This was done by squeezing the foregut contents onto a slide, mounting them in polyvinyl lactophenol, and then examining them with the aid of a compound microscope equipped with a graticule with a 50 × 20 grid, which was used to estimate percentages of food types. Gut contents were divided into 6 food types: vascular plant material (VPM; particles >1 mm), CPOM (particles 50 μm–1 mm), fine particulate organic matter (FPOM; particles <50 μm), algae (ALG), animal material (AM; including whole prey and fragments of exoskeleton), and inorganic materials (IM; mainly sand and silt). For taxa for which no individual's stomach contained food items or when food items were indistinguishable, literature was used to assign an FFG (Dobson et al., 2002; Day and de Moor, 2002a,b,c; de Moor et al., 2003a,b; Merritt et al., 2008).

3.2.4 Community structure and functional composition

The community structure and functional composition were described in terms of relative biomass, numerical abundance, and species richness of all taxa and 4 FFGs (collectors, predators, scrapers, and shredders) that were identified from cluster analysis and literature. Ratios of the various FFGs based on numerical abundance and biomass were calculated and then used as surrogates for ecosystem attributes and for assessing the ecological health of the streams (Vannote et al., 1980; Merritt et al., 2002; Merritt and Cummins, 2006): 1) balance between autotrophy and heterotrophy (production [P]/ respiration [R]) index was calculated as the ratio of scrapers to (shredders + total collectors); 2) linkage between riparian inputs and stream food webs (CPOM/FPOM) was calculated as the ratio of shredders to total collectors; 3) top-down predator control was calculated as the ratio of predators to prey (total of all other groups). Other common ratios, such as the relative dominance of FPOM in suspension compared with that deposited in the benthos and channel stability, were not calculated because details of mouthparts were not examined and, therefore, collectors could not be separated further into gatherers and filterers. No other investigators have published similar studies containing these ratios for African tropical streams, so I based my interpretations of these data on a study done in tropical southern Brazil (Cummins et al., 2005). P/R > 0.75 indicates autotrophy; CPOM/FPOM > 0.25 indicates normal shredder association linked to a functioning riparian zone; and predator/prey between 0.1 and 0.2 indicates a normal predator-to-prey balance, whereas a value >0.2 indicates an overabundance of predators.

3.2.5 Statistical analysis

Two-way analysis of variance (ANOVA) was used to test for differences in physical-habitat, riparian, and organic-matter variables between seasons (dry and wet) and two categories of canopy cover (open and closed), with canopy cover and seasons as main factors and a canopy cover × season interaction term. In cases where canopy cover and season did not influence the dependent variables, one-way ANOVAs was run with stream as the main factor. Habitat conditions expressed as percentages were arcsin($\sqrt{[x/100]}$)-transformed while physicochemical variables, except pH, were $\ln(x + 1)$-transformed before analysis to meet assumptions for parametric tests. Means for the physical-habitat measurements, including stream-size variables and water-quality variables for the 2 seasons were calculated. The species/ morphospecies whose gut contents were examined were assigned to a dietary group according to diet with the aid of cluster analysis (Ward's clustering method; Statistica version 7;

47

StatSoft, Tulsa, Oklahoma) based on average percentages of each food type for all individuals examined (Cheshire et al., 2005). Differences among groups were confirmed with a multivariate analysis of variance (MANOVA) with dietary group as the independent variable and the arcsin($\sqrt{[x/100]}$)-transformed percentages of each food type as the dependent variables. Separate general linear models (GLMs) were then run with single food types as independent variables, followed by Fisher's least significant difference (LSD) *post hoc* tests to identify groups that differed in gut contents for each food type.

The total abundance, biomass, and taxon richness of all taxa, shredders, and nonshredders in litter samples were compared between seasons (dry and wet) and between two categories of canopy cover (open and closed) with 2-way ANOVA with season and canopy cover as the main factors and a season × canopy cover interaction. I used 60% canopy cover as the threshold between open- and closed-canopy sites. Spearman's correlation analysis was used to test for correspondence among macroinvertebrate structural and functional attributes and physical-habitat conditions that represented availability of organic matter (% canopy cover, % leaf litter, litter biomass), water quality (turbidity), and stream size (discharge, stream width and depth). To assess how shredders responded to environmental characteristics in the streams, analysis of covariance (ANCOVA) was used with litter biomass, % canopy cover, % leaf litter, discharge, and stream width and depth as covariates to explore variation in shredder abundance, biomass, and taxon richness between seasons and canopy-cover categories and among streams. Differences were considered as significant at $p < 0.05$ and highly significant at $p < 0.01$ and $p < 0.001$.

3 Results

3.1 Physical-habitat conditions

Season affected DO, discharge, and depth, whereas canopy cover affected temperature, % leaf litter, litter biomass, and turbidity (all $p < 0.05$). Both season and canopy cover influenced water-quality variables, but only canopy cover affected organic-matter characteristics. Closed-canopy sites were in forested catchments where human and livestock activities were limited, whereas open-canopy streams were in agricultural catchments. Most closed-canopy streams were highly shaded (>70% canopy cover), whereas in most open-canopy streams, canopy cover was <50% and discontinuous in sections frequented by livestock and people. Closed-canopy streams were cooler (temperature < 15°C) and had lower turbidity (<60 NTU) than open-canopy sites (Table 3.1). Ranges of conductivity differed between cover types, and were 44.0 ± 3.9 to 97.4 ± 2.3 μS/ cm in closed-canopy streams and 56.7 ± 2.1 to 148.0 ± 2.0 μS/cm in open-canopy streams. However, the means were not significantly different. Season and canopy cover had no interactive effect on any variable tested.

Table 3.1. Elevation and mean (± SE) values for physicochemical characteristics of the study streams. Means are given for variables measured more than once during the dry and wet seasons. % litter = % of substrate covered by leaf litter, [#] indicates closed-canopy streams.

Sites	Altitude (m a.s.l)	Depth (m)	Width (m)	Discharge (m³/s)	Stream order	% canopy cover	% leaf litter	% riffle
Chepkosiom I[#]	2191	0.11 ± 0.05	1.8 ± 0.6	0.01 ± 0.02	1st	85	85	70
Issey I[#]	2146	0.28 ± 0.15	3.1 ± 0.5	0.08 ± 0.03	2nd	70	60	70
Issey II[#]	2087	0.16 ± 0.09	4.1 ± 0.7	0.12 ± 0.04	2nd	65	68	60
Ngatuny[#]	2063	0.25 ± 0.7	4.3 ± 0.5	1.25 ± 0.90	3rd	60	50	65
Philemon[#]	2286	0.08 ± 0.03	2.2 ± 0.2	0.01 ± 0.004	1st	90	80	80
Sambambwet[#]	2096	0.07 ± 0.02	2.1 ± 0.2	0.01 ± 0.005	1st	90	90	55
Chepkosiom II[#]	2060	0.12 ± 0.02	2.5 ± 0.4	0.08 ± 0.01	2nd	80	75	40
Saramek[#]	2091	0.19 ± 0.11	1.9 ± 0.4	0.04 ± 0.12	2nd	80	85	60
Katasiaga[#]	2020	0.19 ± 0.11	1.4 ± 0.4	0.16 ± 0.03	2nd	70	70	65
Mosoriot[#]	2058	0.15 ± 0.05	1.5 ± 0.3	0.12 ± 0.10	2nd	80	80	55
Chepkosiom III	2056	0.12 ± 0.05	3.3 ± 0.3	0.11 ± 0.09	2nd	40	50	50
Issey III	2030	0.21 ± 0.13	2.3 ± 0.7	0.08 ± 0.06	3rd	50	48	40
Issey IV	1982	0.18 ± 0.08	3.7 ± 0.5	0.12 ± 0.10	3rd	50	59	40
Keno	2041	0.07 ± 0.04	3.2 ± 0.4	0.10 ± 0.01	2nd	35	40	55
Mugango	1965	0.23 ± 0.18	5.8 ± 4.2	0.46 ± 0.22	3rd	40	40	55
Nyangores	2043	0.22 ± 0.02	4.6 ± 0.8	0.72 ± 0.30	3rd	45	45	65
Kenjirbei	1979	0.11 ± 0.03	5.5 ± 0.5	0.15 ± 0.12	3rd	45	50	55
Borowet	2050	0.09 ± 0.03	2.1 ± 0.8	0.02 ± 0.13	1st	40	30	55
Tenwek	1953	0.15 ± 0.3	1.3 ± 0.2	0.04 ± 0.02	2nd	50	65	40
Bomet	1911	0.30 ± 0.15	4.6 ± 0.9	0.10 ± 0.10	1st	40	20	55

Table 3.1. Continued

Sites	% pool	pH	DO (mg/L)	Conductivity (µS/cm)	Turbidity (NTU)	Temperature (°C)
Chepkosiom I[#]	30	7.6 ± 0.1	5.7 ± 1.0	97.4 ± 2.3	8.5 ± 1.5	14.6 ± 0.3
Issey I[#]	30	7.2 ± 0.1	8.3 ± 0.3	60.5 ± 3.6	55.3 ± 8.9	14.4 ± 0.3
Issey II[#]	40	7.2 ± 0.1	8.2 ± 0.6	74.2 ± 14.2	56.9 ± 21.3	17.3 ± 3.0
Ngatuny[#]	35	6.5 ± 0.8	7.9 ± 0.3	50.0 ± 2	40.4 ± 15.6	13.6 ± 0.6
Philemon[#]	20	7.3 ± 0.5	6.6 ± 0.5	44.0 ± 3.9	23.4 ± 8.4	12.4 ± 0.3
Sambambwet[#]	45	7.1 ± 0.8	7.7 ± 0.3	91.0 ± 4.2	5.8 ± 1.6	12.2 ± 0.4
Chepkosiom II[#]	60	7.0 ± 0.6	6.8 ± 0.9	75.3 ± 0.3	10.2 ± 2.4	13.4 ± 0.6
Saramek[#]	40	6.9 ± 0.5	7.3 ± 0.4	63.9 ± 0.9	35.6 ± 13.9	13.8 ± 0.8
Katasiaga[#]	35	7.3 ± 0.2	6.9 ± 1.1	72.7 ± 3.2	41.5 ± 13.5	14.1 ± 0.6
Mosoriot[#]	45	7.1 ± 0.2	6.6 ± 1.4	66.3 ± 2.3	9.8 ± 4.8	13.1 ± 0.4
Chepkosiom III	50	6.9 ± 0.3	4.6 ± 2.1	56.7 ± 2.1	159.5 ± 117.5	21.6 ± 0.8
Issey III	60	8.1 ± 0.6	7.0 ± 0.1	123.5 ± 6.5	180.0 ± 56.8	19.2 ± 0.4
Issey IV	60	7.6 ± 0.8	6.4 ± 0.4	148.0 ± 2.0	277.6 ± 56.2	20.6 ± 0.9
Keno	45	7.0 ± 0.6	5.6 ± 0.9	90.2 ± 8.3	124.1 ± 15.9	16.2 ± 1.8
Mugango	45	7.3 ± 0.4	8.3 ± 0.5	89.0 ± 2.3	94.6 ± 27.8	15.6 ± 0.7
Nyangores	35	7.3 ± 0.7	7.5 ± 0.9	58.4 ± 4.4	49.3 ± 29.3	17.5 ± 3.5
Kenjirbei	45	7.4 ± 0.3	7.2 ± 1.3	78 ± 2.3	132.6 ± 23.4	17.2 ± 0.4
Borowet	45	7.1 ± 0.4	5.6 ± 1.2	130.3 ± 5.2	194 ± 56.7	17.6 ± 0.5
Tenwek	60	6.9 ± 0.5	7.2 ± 0.4	75.3 ± 5.9	124.0 ± 30.7	19.3 ± 1.4
Bomet	45	7.1 ± 0.4	6.8 ± 1.5	87.3 ± 9.7	160.2 ± 44.9	19.9 ± 0.2

There were considerable variability among streams. For instance, conductivity and turbidity were variable among individual streams regardless of whether the streams had closed or open canopies (Table 3.1). The lowest DO concentration (4.6 ± 2.1 mg/L) was recorded in an open- canopy stream in an agricultural catchment. The minimum DO in a closed-canopy forested stream was 5.7 ± 1.0 mg/L. pH was similar among streams and did not differ between canopy-cover categories (overall range: 6.5 ± 0.8–8.1 ± 0.6).

3.2 Dietary analysis

Six dietary groups were identified (Figure 3.2). Food items overlapped between groups III and IV and between groups V and VI, but groups differed in their use of food items (MANOVA, Wilk's $\lambda_{30,302} = 0.0002$, $p < 0.0001$). FPOM and CPOM were eaten by 77 and 75 taxa, respectively. Only 41 taxa included animals in their diets, whereas 51 and 46 taxa ate VPM and algae, respectively. A total of 23 taxa (12 predators, 1 shredder, 8 collectors, and 2 scrapers) were classified based on literature because their guts were empty or food items were indistinguishable.

Dietary groups differed in gut contents for every food item (Figure 3.3A–F). Group I mostly contained predators, and an average of 85.5% of their gut contents consisted of animal material. This group had the highest number of taxa (24) and included 7 Diptera, 8 Odonata, 2 Hemiptera, 4 Trichoptera, and 1 Coleoptera, Ephemeroptera, and Plecoptera each. Group II had 14 taxa that were specialist shredders. An average of 74% and 21% of their gut contents were VPM and CPOM, respectively. This group included 7 Trichoptera, 3 Diptera, and 2 Coleoptera and Ephemeroptera. Group III, with 16 taxa, consisted mainly of collectors. An average of 69% of their gut contents was FPOM, 7% was VPM, and 12% was algae. This group included 6 Ephemeroptera, 4 Oligochaeta, 4 Diptera, and 2 Trichoptera. Group IV had 17 taxa and consisted mainly of scrapers. Guts of these taxa contained an average of 36% algae, 34% FPOM, and 21 and 14% inorganic material and CPOM, respectively. This group included 6 Ephemeroptera, 5 Trichoptera, 3 Coleoptera, and 1 Diptera, Gastropoda, and Lepidoptera each. Group V had 9 taxa that were mainly predators and generalist shredders. Their gut contents consisted of 42% animal material, 26% VPM, and 20% CPOM. This group included 5 Diptera, 2 Decapoda (crabs), 1 Coleoptera, and 1 Lepidoptera. Group VI had 6 taxa and consisted of generalist collectors. The group contained 1 Decapoda (shrimp, Atyidae) and 5 Trichoptera (Hydropsychidae). Their guts contained 49% CPOM, 18% FPOM, and 15% VPM.

The allocation of taxa to FFGs was as follows. Predators were all taxa in group I and those in group V whose guts contained >50% animal material. Shredders were all taxa from group II and some taxa of group V (those whose gut contained >50% VPM and CPOM combined). Collectors were species from groups III and VI. Scrapers were all taxa in group IV. Some taxa in group III were classified as collector/scrapers because their guts contained >20% algae. Including invertebrates whose guts were

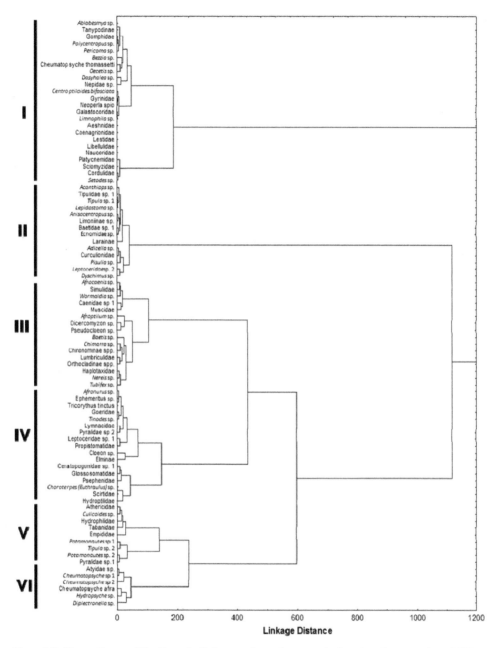

Figure 3.2. Cluster diagram (Ward's method) for macroinvertebrate species based on the proportion of different food types in their guts. Group I consists of predators, group II of specialist shredders, group III of mostly collectors, group IV of mostly scrapers, group V of predators and generalist shredders, and group VI of generalist collectors.

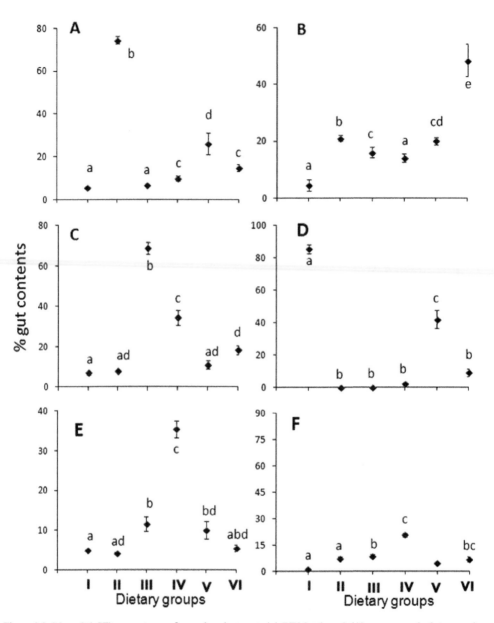

Figure 3.3. Mean (±1 SE) percentages of vascular plant material (VPM; >1 mm) (A), coarse particulate organic matter (CPOM; 50 µm–1 mm) (B), fine particulate organic matter (FPOM; <50 µm) (C), animal material (AM) (D), algae (ALG) (E), and inorganic matter (IM) (F) in the gut contents of individuals belonging to dietary groups I–VI. Points with the same lowercase letter are not significantly different among dietary groups ($p <$ 0.05). See Figure 3.2 for a description of dietary groups.

not analyzed, a total of 44 predators, 27 collectors, 18 scrapers, and 19 shredders were collected in the study area. Shredder taxa included 7 Trichoptera (3 Leptoceridae, 2 Pisuliidae, 1 Calamoceratidae, and 1 Lepidostomatidae), 5 Diptera (4 Tipulidae and 1 Limoniidae:Limoniinae), 2 Ephemeroptera (Baetidae), 2 Coleoptera (1 Elmidae:Larainae and 1 Curculionidae), 2 Decapoda (2 crabs, Potamonautidae) and 1 Lepidoptera (Crambidae); see Plate 3.2 for some examples of shredders in the study area. The involvement of shredders on leaf litter processing was clearly observable in the sampled streams (Plate 3.2G).

3.3 Community structure and functional organization

A total of 20,757 individuals were collected from 109 taxa during the study (81 and 93 taxa during the dry and wet seasons, respectively). Total abundance was higher during the wet than dry season (2-way ANOVA, $F_{1,1} = 11.01$, $p < 0.01$) but did not differ between canopy types. Collectors were numerically dominant during both seasons regardless of canopy cover (Figure 3.4A). Shredder abundance was lower at open- than in closed-canopy sites, whereas scraper abundance was higher at open- than closed-canopy sites during the dry season. *Lepidostoma* sp. was the most widespread and abundant shredder. In closed-canopy sites, its abundance was 10.3 ± 2.4. Crabs were the next most abundant (9.3 ± 5.0), followed by tipulids (3.3 ± 3.4). In open-canopy sites, tipulids were most abundant (13.67 ± 8.67), followed by *Lepidostoma* sp. (6.2 ± 1.6).

Total biomass was higher at closed- than open-canopy sites ($F_{1,1} = 4.16$, $p < 0.05$), but did not differ between seasons. Collectors dominated standing biomass at open-canopy sites during the dry and wet seasons (Figure 3.4B). Shredder biomass was lower at open- than closed-canopy sites, except during the dry season when there was an increase in biomass at open-canopy sites. Scraper biomass was higher at open-canopy sites during the dry than the wet season. Crabs and tipulids contributed >80% of the total biomass at most closed-canopy sites. At open-canopy sites, crabs were rarely encountered, but tipulids occurred at all sites and contributed up to 60% of shredder biomass in the absence of crabs. Large crabs (3536.4 ± 157.2 mg wet mass, up to 56 mm carapace width, Plate 3.2b) and tipulids (59.1 ± 28.9 mg wet mass, up to 54 mm long) were collected in the study area.

Taxon richness was higher at closed- than open-canopy sites ($F_{1,1} = 3.02$, $p < 0.05$) but did not differ between seasons. At open- and closed-canopy sites richness was dominated by predators during the wet season and by collectors during the dry season (Figure 3.4C). The number of collector ($F_{1,1} = 6.39$, $p < 0.05$) and scraper taxa ($F_{1,1} = 9.63$, $p < 0.01$) differed between seasons, and the number of shredder taxa ($F_{1,1} = 8.39$, $p < 0.001$) differed between canopy types. A season × canopy-type interaction affected the number of shredder ($F_{1,1} = 3.94$, $p < 0.05$) and scraper ($F_{1,1} = 2.96$, $p < 0.05$) taxa. During the rainy season, more shredder taxa were found at closed- than open-canopy sites, whereas fewer scraper taxa were found at open- than at closed-canopy sites.

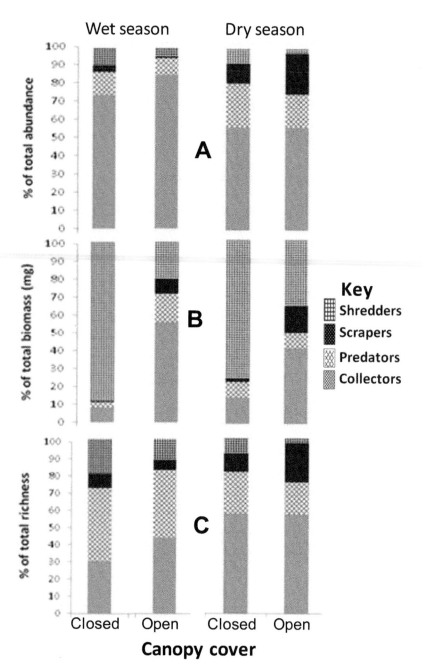

Figure 3.4. Percentage composition of functional feeding groups in terms of relative abundance (A), biomass (B), and species richness (C) for closed- and open-canopy streams during the wet and dry seasons.

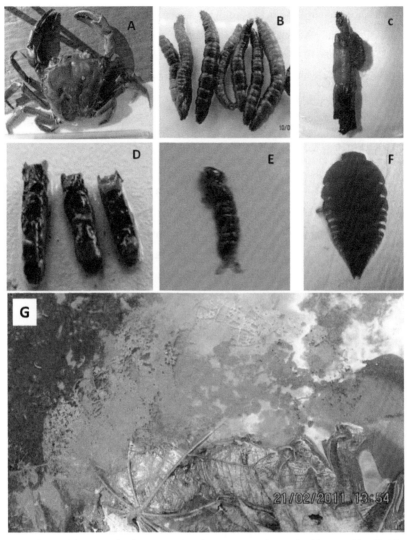

Plate 3.1: Shredders and shredding in the Upper Mara forest streams. Some shredder taxa identified in the study area: A), *Potamonautes* sp., B) *Tipula* sp. 1, C), *Pisulia* sp. D) *Lepidostoma* sp., E) Leptoceridae sp., and F) Larainae sp. G is a leaf of *Croton macrostachyus* with shredder bitemarks.

Macroinvertebrate structural and functional attributes were related to organic matter characteristics (% canopy cover, % leaf litter, and litter biomass), water-quality variables (turbidity and conductivity), and measures of stream size (discharge and width) (Table 3.2). Total abundance (number of individuals/sample) and collectors were favored by high turbidity and conductivity in open-canopy agricultural streams. In contrast, abundances of predators and shredders were negatively related to the same variables but were positively related to % leaf litter, canopy cover, and litter biomass. Scraper abundance was negatively associated with canopy cover, turbidity, and % leaf litter and positively associated with discharge and width.

Shredder abundance, biomass, and richness differed among streams and between canopy types. Litter biomass was significantly associated with shredder richness and marginally related to shredder biomass ($p = 0.08$) (Table 3.3). Season affected shredder biomass and richness, whereas the influence of % leaf litter on shredder biomass and richness was marginally significant ($p = 0.06$ and $p = 0.06$, respectively). Stream size had a minimal effect on shredder distribution and abundance, and discharge only influenced species richness.

3.4 Ecosystem attributes

P/R ratios based on abundance and biomass data indicated that 9 of 10 closed-canopy and 7 of 10 open-canopy streams were heterotrophic (P/R < 0.75; Table 3.4). Use of biomass data yielded stronger indications of heterotrophy than use of abundance data in closed-canopy streams. The effect was reversed in most open-canopy streams. The CPOM/FPOM ratio addressed the integrity of the riparian zone. Abundance data indicated that 6 closed- and 8 open-canopy streams did not have a functioning riparian zone, but biomass data indicated that all streams except one open-canopy stream had a functioning riparian zone (CPOM/FPOM > 0.75). Abundance data indicated that 4 closed- and 5 open-canopy streams had an overabundance of predators (predator/prey > 0.2; Table 3.4). Biomass data indicated that 3 closed- and 5 open-canopy streams had an overabundance of predators.

4. Discussion

4.1 Gut analyses

This study is one of the few studies in the African tropics in which the analysis of gut contents of a large number of stream macroinvertebrates was used to assign taxa to FFGs (Palmer et al., 1993; Dobson et al., 2002). Use of diet rather than morphological and behavioral adaptations for acquiring food to classify macroinvertebrates into FFGs has been questioned because of ontogenic shifts and opportunistic feeding (Palmer et al., 1993 and references therein). Nevertheless, analysis of gut contents enabled me to classify taxa into collector, scraper, shredder, and predator FFGs and has improved understanding of the functional composition and trophic relationships among macroinvertebrates in tropical streams. My work is valuable because growing evidence indicates that taxonomically related species may have different diets in tropical and temperate areas, and discrepancies have been reported when temperate keys have been used to assign tropical-stream invertebrates to FFGs (Yule, 1996; Dobson et al., 2002; Cheshire et al., 2005).

Taxonomic fidelity to particular diets was not evident for most taxonomic groups in this study, and omnivory was prevalent in many taxa. Odonata was the only order that contained specialist predators, although Gomphidae fed on substantial amounts of detritus. The Leptoceridae, represented by specialist shredders elsewhere (Cheshire et al., 2005), included scrapers and predators. The

Table 3.2. Spearman's correlation analysis among macroinvertebrate community attributes and stream characteristics that represent organic matter and riparian conditions (% canopy cover, % leaf litter, and litter biomass), water quality (turbidity and conductivity), and stream size (discharge, depth, and width). Values in bold are significant at $p < 0.05$.

Community attributes	Physical habitat characteristics							
	% canopy cover	% leaf litter	Litter biomass (g/m²)	Turbidity (NTU)	Conductivity (µS/cm)	Discharge (m³/s)	Depth (m)	Width (m)
Total abundance	−0.20	−0.14	0.10	**0.33**	**0.33**	−0.02	0.19	0.01
Shredder abundance	**0.37**	**0.33**	−0.05	−0.21	0.18	0.06	0.02	−0.11
Nonshredder abundance	0.21	0.28	−0.12	0.14	**0.45**	0.00	0.04	−0.15
No. total taxa	0.26	0.22	0.12	−0.22	−0.20	0.17	−0.08	−0.17
No. collectors	**−0.35**	**−0.35**	−0.15	**0.46**	0.12	0.00	0.11	−0.11
No. predators	**0.35**	**0.34**	0.10	**−0.33**	−0.22	−0.08	−0.24	−0.16
No. scrapers	0.02	−0.06	−0.04	**−0.38**	−0.22	**0.36**	−0.18	0.23
No. shredders	**0.40**	**0.37**	**0.32**	**−0.31**	**−0.41**	**−0.33**	0.16	−0.01
% collector individuals	**−0.31**	**−0.41**	−0.14	0.23	**0.43**	0.04	−0.01	0.11
% predator individuals	0.18	0.11	−0.02	**−0.34**	**−0.31**	0.04	−0.18	0.27
% scraper individuals	**−0.39**	−0.28	**−0.31**	0.11	−0.19	**0.37**	−0.01	**0.33**
% shredder individuals	**0.31**	0.24	0.25	**−0.34**	−0.07	0.07	−0.07	−0.10
Collector biomass	−0.30	−0.12	−0.13	0.21	0.16	−0.02	0.01	−0.02
Predator biomass	−0.21	−0.08	−0.01	0.13	0.00	0.15	0.01	0.18
Scraper biomass	−0.30	−0.15	−0.05	**−0.45**	0.05	−0.02	−0.06	0.09
Shredder biomass	**0.34**	0.17	0.13	0.13	−0.13	0.07	−0.01	0.09

Ephemeroptera and Hydropsychidae (Trichoptera), which usually are collector–gatherers/ scrapers and collector–filterers, respectively (Merritt et al., 2008), included predators. *Afroptilum* sp., *Baetis* sp., *Dicercomyzon* sp., and *Pseudocloeon* sp. (Ephemeroptera) showed no preference for a particular food type(s) and were difficult to place into a single FFG. They fed on fine amorphous detritus that could not be easily differentiated, so I had to classify them as either collectors or scrapers. In Madagascar, *Afroptilum* sp. rely more on algal C sources in agricultural streams (where they also have much higher biomass) and amorphous detritus in forested streams (Benstead and Pringle 2004). Omnivory in Crambidae sp. 2, *Potamonautes* sp. 1, *Potamonautes* sp. 2, and *Tipula* sp. 2 indicates that they are facultative detrital shredders. For example, *Potamonautes* spp. consumed high proportions of leaf litter, but they also forage in the riparian zone and adjoining forests to hunt for prey to supplement their diet of leaf litter (Abdallah et al., 2004; Lancaster et al., 2008). During this study, several crabs were observed hiding under rocks and in crevices in the riparian zone, and during the rainy season, a gravid female was captured in the riparian zone, 10 m from the stream.

4.2 Shredder diversity and abundance

Diverse shredder guilds have been reported from tropical areas where few taxa were expected (Cheshire et al., 2005; Yule et al., 2009; Chará-Serna et al., 2012; Salmah et al., 2013). Shredders were diverse (19 taxa) and abundant in closed-canopy forested streams and made up ~17 and 20% of all

Table 3.3. Results of analysis of covariance showing variation of shredder abundance, biomass, and species richness with season and stream and the covariates canopy cover, litter biomass, % leaf litter, discharge, and stream depth and width. * = $p < 0.05$.

Source or variation	df	MS	F	p
Shredder abundance				
Stream	23	38019.8	7.33	**<0.01***
Season	1	34342.6	3.08	0.10
Canopy	1	28582.3	18.62	**<0.01***
Litter biomass	1	20421.4	1.33	0.27
% leaf litter	1	8384.6	0.48	0.49
Discharge	1	10852.9	0.63	0.43
Depth	1	2099.1	0.12	0.73
Width	1	4027.1	0.23	0.63
Error	26	1759.3		
Shredder biomass				
Stream	23	65.5	4.17	**0.03***
Season	1	71.6	5.45	**0.03***
Canopy	1	75.8	15.69	**0.02***
Litter biomass	1	40.1	3.05	0.08
% leaf litter	1	45.6	3.67	0.06
Discharge	1	37.8	2.87	0.10
Depth	1	24.5	1.86	0.18
Width	1	6.3	0.48	0.49
Error	26	13.14		
Shredder species richness				
Stream	23	7.3	6.43	**0.02***
Season	1	9.5	8.41	**0.01***
Canopy	1	27.1	23.94	**<0.01***
Litter biomass	1	11.3	9.49	**<0.01***
% leaf litter	1	4.4	3.99	0.06
Discharge	1	32.2	28.43	**<0.01***
Depth	1	0.4	0.38	0.54
Width	1	0.2	0.14	0.71
Error	26	7.51		

taxa, and 75 and 84% of total macroinvertebrate biomass during the dry and wet seasons, respectively. The 19 shredder taxa encountered in this study compare well with a number of temperate and tropical streams that are not shredder limited (Table 3.5). However, the findings differ strongly from results of previous studies in eastern Africa in which, according to Tumwesigye et al. (2000), Dobson et al. (2002), Abdallah et al. (2004), and Masese et al. (2009b), 6% of all individuals, 17% (5 of 36 taxa), 7% (3 of 41 taxa), and 11% of taxa (6 of 56 taxa), respectively, were shredders. I see 4 possible explanations for the large number of shredder taxa recorded in my study. First, earlier investigators placed taxa into FFGs based solely on literature for temperate streams, except that Dobson et al. (2002) also examined mouthparts of 44% and gut contents of 11% of taxa collected. Dobson et al. (2002) noted that they may have allocated taxa to FFGs incorrectly when they found that a tropical African baetid, *Acanthiops* sp., was a shredder (baetids in northern temperate streams are usually scrapers or collector–gatherers; e.g., Merritt et al., 2008). This kind of discrepancy suggests that tropical shredders may be overlooked (misclassified) when temperate- stream keys are used (Camacho et al., 2009). Second, the coarse taxonomic resolution (mostly to family) used by Dobson et al. (2002)

Table 3.4. Mean values of stream ecosystem attributes derived from ratios of macroinvertebrate functional feeding groups in the upper Mara River basin, Kenya. Ratios are based on numerical abundance and biomass of functional feeding groups (FFGs). The autotrophy to heterotrophy (production [P]/respiration [R]) index was calculated as the ratio of scrapers to (shredders + total collectors). Coarse particulate organic matter (CPOM)/fine particulate organic matter (FPOM) is the ratio of shredders to total collectors. Top-down predator control is the ratio of predators to all other FFGs. [#] indicates closed-canopy streams. Boldface indicates autotrophy, functioning riparian zone, or an over-abundance of predators with a strong top-down control.

| | Ecosystem attributes | | | | | |
| | Abundance | | | Biomass | | |
Sites	P/R	CPOM/FPOM	Top-down predator control	P/R	CPOM/FPOM	Top-down predator control
Chepkosiom I[#]	0.13	0.14	**0.22**	0.59	**1.05**	**0.39**
Issey I[#]	0.52	**0.29**	0.12	0.01	**36.69**	0.08
Issey II[#]	0.27	0.05	0.05	0.03	**20.59**	0.01
Ngatuny[#]	**1.61**	**0.37**	0.18	0.07	**13.65**	0.03
Philemon[#]	0.49	**0.29**	**0.31**	0.25	**0.26**	0.10
Sambambwet[#]	0.25	**0.33**	**0.23**	0.03	**11.65**	**0.25**
Chepkosiom II[#]	0.52	0.14	**0.29**	0.31	**3.65**	**0.26**
Saramek[#]	0.28	0.21	0.19	0.01	**20.02**	0.03
Katasiaga[#]	0.51	0.12	0.18	0.05	**19.82**	0.04
Mosoriot[#]	0.37	0.17	0.08	0.06	**4.55**	0.15
Chepkosiom III	0.62	0.03	0.13	**1.35**	0.87	0.09
Issey III	0.38	0.01	0.11	0.56	0.03	**0.35**
Issey IV	0.15	0.01	0.03	0.23	0.47	0.15
Keno	**0.97**	**0.40**	**0.31**	0.04	**32.89**	0.02
Mugango	**1.28**	**0.43**	0.05	0.58	**6.91**	0.02
Nyangores	**1.53**	0.09	**0.25**	**0.85**	**5.02**	**1.00**
Kenjirbei	0.43	0.05	**0.23**	0.19	0.57	**1.06**
Borowet	0.48	0.09	**0.40**	**0.85**	1.76	**0.29**
Tenwek	0.20	0.01	0.01	0.29	0.37	0.14
Bomet	0.68	0.10	**0.37**	**0.79**	0.29	**3.00**

also contributed to the low number of shredder taxa identified. Examples in Dobson et al. (2002) include a shredder, Larainae sp., which was classified with scrapers as part of Elmidae, 3 leptocerids and 4 tipulids that were each grouped under 1 family as Leptoceridae and Tipulidae, respectively. Third, the abundance and distribution of shredders in tropical streams can be temporally and spatially variable (Cheshire et al., 2005; Camacho et al., 2009), and the study by Dobson et al. (2002) did not consider seasonal changes. The potential for higher diversity and abundance of macroinvertebrate taxa during the rainy than the dry season cannot be ignored. For instance, as part of their life-history strategies, some tropical insects mature and emerge during the rainy season (Mathooko, 2001; Jacobsen et al., 2008). Moreover, water quality in some streams deteriorates during the dry season (Masese et al., 2009a). Most shredder taxa in this study occurred in low abundance, so a seasonal sampling scheme was needed to avoid missing important taxa. For example, 4 of the 19 shredders identified were collected only during the rainy season. Last, the sampling method was adjusted to capture large invertebrates that could avoid capture by standard sampling methods. The rapid sampling method enabled capture of many crabs in closed-canopy streams where they were most prevalent. Crabs (*Potamonautes* spp.) occurred in some streams sampled by Dobson et al. (2002), but many were

Table 3.5. Comparison of the number of shredder taxa among aquatic invertebrates across low order (1st-3rd order) streams from different regions and climates around the world. The number of shredders taxa include obligate and facultative (in brackets) shredders in studies with separation between the 2 groups. Taxa assigned partially to shredders (multiple functional feeding groups) were considered as facultative shredders.

Region/ stream or catchment	Country	Climate	Number of shredder taxa	Remarks	References
Mara streams	Kenya	Tropical	19	10 closed-canopy streams	Present study
Mara streams	Kenya	Tropical	12	10 open-canopy streams	Present study
Central highlands	Kenya	Tropical	5	Forest and open canopy streams	Dobson et al. 2002
Bougainville island	Papua New Guinea	Tropical	7	-	Yule 1996
Queensland	Australia	Tropical	10 (4)	Forested upland	Cheshire et al. 2005
Malaysian streams	Malaysia	Tropical	22	12 upland streams	Yule et al. 2009
Austaralian Wet Tropics	Australia	Tropical	12	Six streams	Camacho et al. 2009
Panamanian streams	Panama	Tropical	16	Six streams	Camacho et al. 2009
Otun River basin	Colombia	Tropical	9	3 forested streams	Chara-Serna et al. 2011
Hong Kong	China	Tropical	6 (2)	5 shaded streams	Li and Dudgeon 2008
Hong Kong	China	Tropical	3 (1)	5 unshaded streams	Li and Dudgeon 2008
Belum	Malaysia	Tropical	12	Forested catchment	Salmah et al. 2013
Berembun	Malaysia	Tropical	13	Forested catchment	Salmah et al. 2013
Gunung Angsi	Malaysia	Tropical	13	Forested catchment	Salmah et al. 2013
Gunung Tebu	Malaysia	Tropical	10	Forested catchment	Salmah et al. 2013
Hulu Gombak	Malaysia	Tropical	15	Forested catchment	Salmah et al. 2013
Keledang Saiong	Malaysia	Tropical	15	Forested catchment	Salmah et al. 2013
Sarawak	Borneo	Tropical	14	Forested catchments	Jinggut et al. 2012
Mambucaba River Basin	Brazil	Tropical	12 (5)	Forested catchments	De Oliveira and Nessimian 2010
Argentina	Argentina	Temperate	14	-	Miserendino and Pizzolon 2004
North Carolina	United States	Temperate	13	-	Stone and Wallace 1998
New Hampshire	United States	Temperate	14	-	Hall et al. 2001
West Virginia	United States	Temperate	19	-	Angradi 1996
Southwest Ireland	Ireland	Temperate	16	-	Murphy and Giller 2000
Southeast England	England	Temperate	10	-	Dobson and Hildrew 1992
Northeast France	France	Temperate	25	-	Dangles and Guérold 2000
Southeast France	France	Temperate	12	Nutrient enriched	Lecerf et al. 2006
Finnish streams	Finland	Temperate	17	-	Heino 2009

not captured by the method they used. Later studies have revealed high density and biomass of crabs in Kenyan highland streams (Dobson et al., 2007b; Magana et al., 2012).

Most of the shredder taxa identified in this study are represented in temperate streams. Baetids (Ephemeroptera), crabs (Potamonautidae), and Pisuliidae (Trichoptera) are the exceptions. The 2 baetid shredders so far identified in Kenyan streams (Dobson et al., 2002, this study), the genera *Barba* (Leptophlebiidae) from Papua New Guinea (Yule, 1996), and *Atalophlebia* (Leptophlebiidae) from the Australian tropics (Cheshire et al., 2005) are shredders, whereas members of these groups are mostly collector–gatherers or scrapers in temperate streams (e.g., Merritt et al., 2008). Crabs are widespread in African streams where they display high endemism (Dobson, 2004) and are responsible for rapid breakdown of leaf litter (Moss, 2007). On the other hand, key shredder taxonomic groups in temperate streams were either poorly represented or absent in my streams. Plecoptera, which is diverse with many shredder species in temperate streams, was depauperate in my streams and was represented by only 1 species (*Neoperla spio*, Perlidae), which is a predator. Decapod shrimps are major shredders in temperate and some tropical streams, but the single Atyidae species in my streams is a collector. Common shredder taxa in temperate streams that have been found in other tropical streams, but not in my study, include Limnephilidae, Sericostomatidae, Peltoperlidae, Leuctridae, and Nemouridae (Cheshire et al., 2005; Yule et al., 2009; Salmah et al., 2013).

Most remarkable were the large sizes, high densities, and high biomasses of *Potamonautes* spp. and *Tipula* spp. in closed- and open-canopy streams, respectively. Large size (Plate 3.2) seems to be common feature of *Potamonautes* spp. in east African highland streams (Abdallah et al., 2004; Dobson et al., 2007a,b; Moss, 2007) and is important for the breakdown of recalcitrant leaves in tropical streams (Wantzen and Wagner, 2006; Moss, 2007; Yule et al., 2009). No detailed analysis has been done of nutritional quality of riparian vegetation in highland streams in Kenya, but most tree species have waxy leaves that are not favorable to shredders (Dobson et al., 2002).

4.3 Physical-habitat effects

Season influenced structural and functional organization of macroinvertebrates by accentuating differences in water quality and habitat characteristics. The abundance of most taxa was considerably lower in the dry than in the wet season (see also Harrison and Hynes, 1988; Mathooko and Mavuti, 1992; Masese et al., 2009a). However, in other studies of tropical streams, abundance increased during the dry season (Tumwesigye et al., 2000; Arimoro et al., 2012). Flow reduction during the dry season contributes to seasonal variability in physicochemical conditions that could influence macroinvertebrate communities. For instance, I recorded the lowest DO and highest conductivity during the dry season in open-canopy streams. These streams were in areas frequented by people and livestock and were subject to sedimentation and input of nutrients and organic wastes by runoff. Thus, algal food sources for scrapers probably were smothered during the wet season. An increase in the number of scrapers in these streams during the dry season probably was related to increased algal availability as a result of reduced turbidity.

Canopy cover strongly influenced the distribution and abundance of scrapers and shredders. Nine of the 19 shredder taxa found during my study occurred only in closed-canopy streams. Crabs do occur in open-canopy streams in agricultural areas, but their abundance is very low (occasionally 1 or 2 mature individuals were captured). The importance of canopy cover is supported by observations that reproduction in crabs is more successful in forest streams which serve as nurseries from which adults migrate downstream (Dobson et al., 2007a). Changes in water temperature, conductivity, and litter input probably contributed to the skewed occurrence of shredder taxa in closed-canopy streams. Open-canopy streams were warmer and leaf-litter input was lower and dominated by the exotic *Eucalyptus* spp. than closed-canopy streams where riparian vegetation was speciose. Water temperature and leaf-litter characteristics are globally important factors affecting shredder abundance and diversity (Boyero et al., 2011a,b,c).

Difference in stream size as determined by discharge, width and depth were a reflection of human activities both at the reach and catchment scales. Open-canopy streams were flashy resulting in higher discharge levels during spates. In-stream human and livestock activities also widened channels in open-canopy streams. These human- and livestock-induced changes in stream size are different from the natural increases in stream order whose effects on FFGs were not evident in the 1st-3rd order streams considered in this study.

4.4 Ecosystem attributes

I used the ratio of scrapers (shredders + total collectors) as a surrogate for P/R, and applied thresholds proposed by Cummins et al. (2005) for tropical Brazilian streams to the P/R ratios in streams in this study. Most streams were heterotrophic, but open-canopy streams in agricultural areas tended to be autotrophic. Some open-canopy streams, such as Issey III, Issey IV, and Tenwek, receive organic pollution from livestock. These streams had high abundances of collectors (oligochaetes and Chironominae), which shifted abundance-based P/R ratios in these potentially autotrophic streams toward heterotrophy. Most shredders in the study area, such as crabs, tipulids, and trichopterans (*Pisulia* sp. and *Lepidostoma* sp.), were large. The presence of large shredders, especially crabs (*Potamonautes* spp.), in closed-canopy and in some open-canopy agricultural streams shifted biomass-based P/R ratios toward greater heterotrophy.

The influence of shredder body size also was evident in the CPOM/FPOM ratio that addresses the integrity of the riparian zone. When biomass data were used, all closed-canopy streams and 9 of 10 open-canopy streams passed the criteria for a functioning riparian zone. However, when abundance data were used, only 4 of the closed- and 2 of the open-canopy streams passed this criterion. These results raise the issue of potential bias in biomass-based surrogates for measures of ecosystem functioning when large-bodied shredders are dominant. Assessments based on abundance-based surrogate ratios reflected assessments based on visual evidence of impacted riparian zones and removal of indigenous vegetation along open-canopy streams in agricultural areas. Closed-canopy streams are not spared from human influences, and it is common to find domestic animals grazing and selective cutting of trees for timber and firewood (Minaya et al., 2013). Thus, the uniformly positive biomass-based assessments of riparian-zone integrity in closed-canopy streams are suspect.

High abundance of predators was evidence of strong top-down control in some open canopy streams, such as Bomet and Borowet, where flow decreased considerably during the dry season and pools were dominated by predacious families, such as Notonectidae (backswimmers) and Lestidae. Fewer closed-canopy streams had an overabundance of predators, but this assessment probably would change if omnivorous crabs were included in the predator category. Crabs exert top-down controls on other macroinvertebrates and take part in rapid breakdown of leaf litter in other streams in the region (Moss, 2007; Lancaster et al., 2008).

Overall, the FFG ratios provided evidence of widespread human influences in the study area. Examination of mouth parts to separate collectors into filterers and gatherers will provide more rigorous and complete criteria for assessing the effects of riparian disturbances, sedimentation, and the quality of FPOM transported by these streams (Merritt and Cummins, 2006). The threshold values for heterotrophy vs autotrophy should be re-examined in these streams, given that very few studies are available in the tropics for comparison, and the role of livestock should be investigated further.

Conclusions

The task of assigning stream macroinvertebrates to FFGs is not straightforward, and is, at times, controversial, especially when assignments are not supported by details on feeding modes and the structure of mouthparts. Some taxa have too variable a diet to be assigned to a FFG, and some taxa undergo ontogenic shifts in diet. Nevertheless, gut-content analysis enabled me to classify macroinvertebrates collected in this study into FFGs and to contribute information to the growing data on the functional organization of tropical streams, including dietary requirements and trophic relationships. The FFG ratios offered a glimpse into the overall functioning of these streams and reflected a shift from heterotrophy to autotrophy arising from changing land use and clearing of riparian vegetation. This study highlights the importance of shredders in these streams and the effects of riparian alterations on invertebrate community composition and ecosystem functioning. The wider consequence of these effects on ecosystem services merits further research and serious consideration when planning future landuse changes in eastern Africa.

Chapter

4

Litter processing and shredder distribution as indicators of riparian and catchment influences on ecological health of tropical streams

Publication based on this chapter:
Masese FO, Kitaka N, Kipkemboi J, Gettel GM, Irvine K, McClain ME. 2014. Litter processing and shredder distribution as indicators of riparian and catchment influences on ecological health of tropical streams. Ecological Indicators 46: 23–37.

Abstract

Terrestrial plant litter is the main source of energy for food webs in forest headwater streams. Leaf litter quality often changes when native tree species are replaced by exotic ones and land use change in the watershed can alter physico-chemistry and functional composition of invertebrate communities, ultimately impairing associated ecosystem processes. The composition of invertebrate functional feeding groups (FFGs) and the ecosystem process of leaf breakdown were used as structural and functional indicators, respectively, of ecosystem health in upland Kenyan streams. During dry and wet conditions, invertebrates were sampled in 24 streams within forest (10), mixed (7) and agriculture (7) catchments. Five forest and five agriculture streams were subsequently used for leaf litter breakdown experiments using two native (*Croton macrostachyus* and *Syzygium cordatum*) and one exotic (*Eucalyptus globulus*) species differing in quality. Coarse- and fine-mesh litterbags were used to compare microbial (fine-mesh) with shredder + microbial (coarse-mesh) breakdown rates, and by extension, determine the role of shredders in litter processing in these streams. Seasonal influences on water quality were observed across catchment land uses. Total suspended sediments, turbidity and total dissolved nitrogen were consistently higher during the wet than dry season. However, seasonal influences on FFGs were inconsistent. Catchment land use influenced invertebrate functional composition: 21 taxa, including eight shredders, were restricted to forest streams, but abundance was a poor discriminator of disturbance. Breakdown rates were generally higher in coarse- compared with fine-mesh litterbags for the native leaf species and the relative differences in breakdown rates among leaf species remained unaltered in both agriculture and forest streams. Shredder and microbial breakdown of leaf litter displayed contrasting responses with shredders relatively more important at forest compared with agriculture streams. However, these patterns were inconsistent across leaf species over the dry and wet seasons. Overall, shredder mediated leaf litter breakdown was dependent on leaf species, and was highest for *C. macrostachyus* and lowest for *E. globulus*. This suggests that replacement of indigenous riparian vegetation with poorer quality *Eucalyptus* species has the potential to reduce nutrient cycling in streams, with foodwebs becoming more reliant on microbial processing of leaf litter, which cannot support diverse consumers and complex food webs.

1. Introduction

Developing landscapes to meet human needs has altered surface water hydrology, geomorphology and physico-chemistry, impacting the ecology of streams (Allan, 2004; Dudgeon et al., 2006; Vorosmarty et al., 2010). Land-use changes in catchments and along riparian corridors have replaced natural forests with agriculture, pastures and exotic forestry species (Ferreira et al., 2006; Hladyz et al., 2011). Loss of natural riparian corridors alters stream light and temperature regimes and the timing, quality and quantity of inputs of leaf litter and dead wood (Elosegi and Johnson, 2003; Wantzen et al., 2008), in turn increasing nutrient inputs and primary production (Scarsbrook and Halliday, 1999; Baxter et al., 2005). These changes typically reduce habitat complexity and biodiversity, and affect organic matter dynamics, nutrient cycling, water purification and erosion processes (Palmer and Filoso, 2009; Acuña et al., 2013).

Riverine ecosystems exhibit extreme heterogeneity in environmental conditions at multiple temporal and spatial scales ranging from microhabitats to whole landscapes (Frissell et al., 1986; Poff et al., 1997). Tropical streams and rivers are highly dynamic and water quality is influenced by both catchment land use, and riparian and in-stream activities (Jinggut et al., 2012; Minaya et al., 2013; Silva-Junior et al., 2014). For example, elevated concentrations of nutrients and sediments have been recorded in streams draining agricultural catchments during the wet season due to run-off from unpaved roads, footpaths and farmlands (Kilonzo et al., 2013). In rural catchments, in-stream human activities (water abstraction, bathing, washing and watering of livestock) are influenced by weather conditions, being more common during the dry season (Mathooko, 2001; Yillia et al., 2008). The quality and quantity of leaf litter inputs into these streams is also seasonally variable and dependent on catchment and riparian conditions (Wantzen et al., 2008).

Assessing anthropogenic disturbances on streams relies mostly on monitoring metrics of aquatic communities and physico-chemistry (Barbour et al., 1999; Bonada et al., 2006). The relative abundances of various taxa and functional feeding groups (FFGs) of stream benthic invertebrates have been used as structural indicators (Rosenberg and Resh, 1993; Barbour et al., 1999), while functional components (Gessner and Chauvet, 2002; Young et al., 2008) have tended to be neglected. Structural and functional indicators are not necessarily concordant, highlighting the need to consider both during bioassessment (Gessner and Chauvet, 2002; Bonada et al., 2006). Breakdown of leaf litter is an important functional indicator that links riparian vegetation, environmental conditions, microbial and invertebrate activities (Vannote et al., 1980; Hladyz et al. 2010; Woodward et al., 2012). However, while the use of leaf litter breakdown as a measure of ecosystem functioning is receiving increased attention in temperate streams (Gessner and Chauvet 2002; Young et al., 2008; Woodward et al., 2012), studies in the tropics are limited (Jinggut et al., 2012; Silva-Junior et al., 2014), and the influence of seasonality is not well understood. Moreover, variability of leaf litter decomposition is likely across climatic regions that differ in environmental factors and invertebrate functional composition (Pozo et al., 2011).

Despite under-representation in the literature, the functioning of freshwater tropical ecosystems is highly impacted by human disturbance (Dudgeon et al., 2006). For Africa, balancing increased demands for economies, food production, clean water and environmental quality is an increasing challenge (McClain, 2013; McClain et al., 2014). Many catchments in East African montane forests and high potential agro-ecological zones have lost extensive areas of native vegetation to exotic forests, farming, settlement and grazing (Mati et al., 2008; Maitama et al., 2009). Many landscapes are criss-crossed with unpaved roads and footpaths with extensive areas under grazing and farmlands of mainly fast maturing crops. This risks increased soil erosion and sediment loading to streams and rivers; most notably during the rainy season. Indigenous vegetation along streams and rivers in agriculture catchments is increasingly replaced by exotic *Eucalyptus* species. Even though *Eucalyptus* spp. belong to the same group (Family:Myrtaceae) as *Syzygium cordatum* that is endemic and dominant along riparian areas (Mathooko and Kariuki, 2000), *Eucalyptus* leaves are highly sclerophyllous (Graça et al., 2002). As a result, most leaf litter that enters agriculture streams is refractory and high in polyphenolic compounds. This can significantly alter microbial, fungal and invertebrate communities that colonize leaves, leaf-litter breakdown rates, and the higher trophic levels supported by allochthonous resources (Graça et al., 2002; Ardón and Pringle, 2008). Partitioning effects of pressures on stream biota at different spatial and temporal scales is often lacking, but necessary to guide management and safeguard ecosystem services. With few exceptions, data on ecosystem functioning and the extent of anthropogenic influences on East African streams remain limited (Masese and McClain, 2012).

Leaf litter breakdown in streams is driven by resource quality, activity of consumers and environmental conditions (Tank et al., 2010). The chemical composition and physical structure of leaf litter influence preferences of shredders and microbial colonization rates (Graça et al., 2001; Ligeiro et al., 2010). Nutrient enrichment of streams can accelerate leaf litter breakdown by stimulating microbial activities and invertebrate consumption (Rosemond et al., 2002; Gulis and Suberkropp, 2003), but the stimulation effect is also dependent upon the quality of leaf litter (Ferreira et al., 2006; Gulis et al., 2006). However, land use influences on the diversity and abundance of shredders, which are more sensitive to nutrient pollution compared with microbes (Hieber and Gessner, 2002), imply that contrasting responses to resource quality among microbes and shredders might be expected. In the tropics, higher temperatures stimulate fast rates of microbial breakdown of litter and may reduce food availability for shredders (Irons et al., 1994; Boyero et al., 2011c). Despite the narrow temperature range in the tropics, agriculture streams are warmer and with higher electrical conductivity, suspended sediments and dissolved nutrients compared with forest streams (Kasangaki et al., 2008; Kilonzo et al., 2013; Minaya et al., 2013). Effects of these changes on ecosystem functioning are poorly understood in African tropical streams where land use change has been linked to terrestrial biodiversity loss and changes in the natural flow regimes of rivers (Maitama et al., 2009; Mango et al., 2011; McClain et al., 2014).

In this study, I compared the use of functional and structural indicators to detect changes in land use. I used leaf litter breakdown as a functional indicator and the composition of invertebrates FFGs as structural indicators to assess the influence of rural land use and riparian conditions on ecosystem functioning of Kenyan highland streams. Wet and dry seasons functional organization of benthic invertebrates were characterized in 24 streams distributed among forest, mixed and agriculture catchments. Leaf breakdown experiments were conducted using two native (*Croton macrostachyus* and *Syzygium cordatum*) and one exotic (*Eucalyptus globulus*) leaf species in five forest and five agriculture streams. The two native species represent nearly the extremes of the litter quality range of native riparian trees in the study area, and the exotic species used has lower litter quality and is common along farm edges, and planted as woodlots in riparian areas as well as in commercial plantations. I hypothesize that: 1) the influence of land use and riparian disturbance on invertebrate structural and functional organization are accentuated by seasonality, 2) breakdown rates of all plant species will be higher in agriculture streams than in forest streams, 3) the relative importance of invertebrates shredders on litter breakdown will be lower in agricultural than in forest streams, and 4) breakdown rates will be species-specific with the nutrient poor species displaying the slowest response to land use.

2. Materials and Methods

2.1 Study area

The study was conducted in mid-altitude (1900 m – 2300 m a.s.l) first to third order streams draining the western slopes of the Mau Escarpment within the Kenyan Rift Valley. A total of 24 sites were selected in the headwaters of the Mara River, which flows to Lake Victoria. The river drains the extensive tropical moist broadleaf Mau Forest Complex (MFC) that is a source of rivers draining into Lakes Baringo, Nakuru and Victoria (Figure 4.1). Vegetation in the MFC is diverse, with over 95 tree species (Blackett, 1994). Catchment and sub-catchments were delineated and land use categorized using a combined digital elevation model (DEM), remote-sensing images (Landsat 5 Thematic Mapper data of 2008, 30 m resolution) and topographic maps (1:50,000 survey of Kenya 1971). The area of each land use type for each sub-catchment draining to a sampling site (point) was calculated. Sites were selected in one of 3 catchment-scale land uses that were defined as: a) forest sites (FOR, n = 10) draining catchments with >60% forest; b) agriculture sites (AGR, n = 7) draining catchments with >60% agriculture; and c), mixed sites (MIX, n = 7) located in agricultural areas with the upstream catchments comprising different proportions of the two main land uses, forestry and agriculture, but none with >60%. FOR sites had a mean (±SD) value of 95.3±13.4 % of total catchment area under forestry with other land uses (grasslands, shrubland) comprising <10% (Table 1). All FOR sites did not have any proportion of agricultural land use. AGR sites had a mean (±SD) of 93.8±3.4% of total catchment area under agriculture with the rest of the land uses (roads, bare ground, urban areas) comprising <10% of the land area. All AGR sites did not have any proportion of forest land use. MIX sites had means of 51.4±11.4% and 44.0±8.6% of land area under forestry and agriculture, respectively. Among the three land uses, sites were selected based on accessibility, stream size (discharge), in-stream gradient and physical habitat conditions.

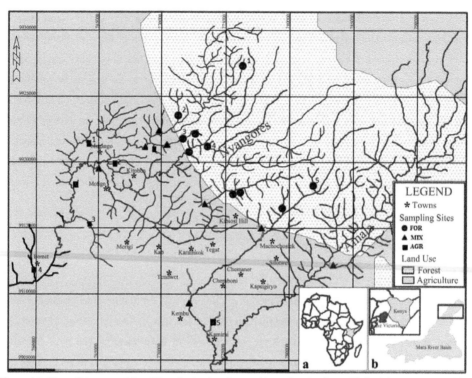

Figure 4.1. Map of the upper Mara River basin showing position of study sites. All the 24 streams were kick-sampled for invertebrates while the numbered streams 1-5 in the agriculture (darker shading) and forest (lighter shading) land uses were used for leaf litter processing experiments. Modified from Minaya et al., 2013.

The present coverage of the MFC is fragmented and reduced in size because of excisions for human settlement, coniferous forest plantations and large- and small-scale cultivation of tea (Lovett and Wasser, 1993). Some intact forest blocks are protected as part of forest reserves and national parks (Lovett and Wasser, 1993). People living in the adjoining areas are involved in semi-intensive smallholder agriculture, characterized by cash crops (mainly tea), food crops (mainly maize, beans and potatoes) and animal husbandry. This has also resulted in the loss of indigenous riparian vegetation along agriculture streams and rivers where exotic *Eucalyptus* dominate.

Climate of the MFC is characterized by low ambient temperatures, falling below $10^\circ C$ during the cold months of January-February. Annual precipitation ranges from 1000 - 2000 mm and is bimodal. Dry conditions occur during January-March and July and two wet conditions during April-June and October-November, which are periods for the long and short rains, respectively.

Table 4.1. Mean (±SE) values for different proportions of land uses in the Agriculture, Mixed and Forest streams in the upper Mara River basin, Kenya; agriculture - n = 7, mixed - n = 7, forest - n = 10.

Land use proportions	Stream types		
	Agriculture	Mixed	Forest
% Forest	-	51.4±11.4	95.3±13.4
% Agriculture	93.8±3.4	44.0±8.6	-
% Other land uses	6.3±4.2	4.6±1.2	5.8±3.3

2.2 Physical and chemical variables

At each site, percent canopy cover, stream width, water depth, velocity and discharge were determined over a 100m reach. Percent canopy cover above each stream was estimated visually. Stream width was measured at 10 transects located at mid-points of 10-m intervals. On each transect, water depth was measured at least at 5 points across the river using a 1-m ruler. Velocity was measured at the same points as depth using a mechanical flow meter (General Oceanic 2030). Stream discharge was estimated using the velocity-area method (Wetzel and Likens, 2000). Presence or absence of leaf litter was noted at each point where depth measurements were made and used to estimate percentage of substratum covered by leaf litter. The proportion of riffles and pools in the stream was determined by identifying whether each of the 10 transects crossed a pool or a riffle, and recorded as a percentage of the 100-m reach. Concurrent measurements of pH, dissolved oxygen (DO), temperature and electrical conductivity were measured *in situ* using a YSI multi-probe water quality meter (556 MPS, Yellow Springs Instruments, Ohio, USA), and turbidity was measured using a portable Hach turbidity meter (Hach Company, 2100P ISO Turbidimeter, USA). Water samples were collected from the thalweg using acid washed HDP bottles for analysis of nutrients, major anions and cations, dissolved organic carbon (DOC) and particulate organic matter (POM). For total suspended solids (TSS) and POM stream water samples were immediately filtered through pre-weighed glass-fibre filters (Whatman GF/F, pre-combusted at 450°C, 4 h). GF/F filters holding suspended matter were carefully folded and wrapped in aluminium foil before transport in a cooler box at 4 °C to the laboratory. Both the filtered and unfiltered water samples were stored and transported in a cooler box and frozen within 10 hours of sampling.

Water quality variables determined in the laboratory were alkalinity, TSS, POM, DOC, total dissolved nitrogen (TDN), total dissolved phosphorus (TDP), total phosphorus (TP), soluble reactive phosphorus (SRP) and total nitrogen (TN). Alkalinity (mmol l^{-1}) was determined by potentiometric titration of 200 ml of filtrate with 0.1 N HCl acid (3.65 g l^{-1}) to a pH of 4.3. GF/F filters holding suspended matter were dried (95°C) to constant weight and TSS was determined by re-weighing on an analytical balance and subtracting the filter weight. Filters were ashed at 500°C for 4 h and re-weighed for determination of POM as the difference between TSS and ash-free-dry weight. DOC and TDN concentrations were determined using a Shimadzu TOC-V-CPN with a coupled total nitrogen analyzer unit (TNM-1). TDP was determined using the ascorbic acid spectrophotometric method, while TP, SRP NH_4-N and TN were determined using standard calorimetric methods (APHA, 1998). Major anions, nitrate (NO_3^-), ortho-phosphate (PO_4^{3-}), chloride (Cl^-) and sulphate (SO_4^{2-}), were determined by Ion Chromatography (Dionex ICS-1000) and the major cations, sodium (Na^+), potassium (K^+), calcium (Ca^{2+}), magnesium (Mg^{2+}), dissolved silicates (DSi), using an ICP-MS. Organic matter in kick samples was washed into a 100-μm-mesh sieve to remove inorganic materials. This size of organic matter that is > 100 μm is collectively referred to as particulate organic matter (POM) in this study. POM was dried to a constant mass at 68°C for at least 48 h, and the different fractions (leaves, sticks, seeds and flowers) weighed separately using a Sartorius balance.

2.3 Leaf litter breakdown

Leaf litter breakdown experiments were conducted during wet- (May-July 2011) and dry- (January-March 2012) seasons at five forest streams and five agriculture streams in rural agriculture catchments. These ten sites were part of the 24 sites for invertebrate kick sampling, and were a compromise between representativeness of environmental conditions and effort.

Three different leaf species were used to test the influence of litter quality on invertebrate colonisation and breakdown. I chose two indigenous leaf species- Lace-leaf *Croton macrostachyus* Hochst. ex Delile (Family Euphorbiaceae) [henceforth *Croton*] and *Syzygium cordatum* Hochst ex Krauss (Family Myrtaceae) [henceforth *Syzygium*]- that are typical riparian trees along streams in the region and are among the ten most dominant species in the MFC (Blackett, 1994; Mathooko and Kariuki, 2000), and one exotic species -*Eucalyptus globulus* Labill. (Family Myrtaceae) [henceforth *Eucalyptus*]- which is the most common replacement species along streams in agriculture catchments in the region (Kenya Forestry Service, 2009). *Croton* leaves are soft and fast decaying and thus were chosen to represent high quality leaf litter in the region. In contrast, *Syzygium* has tough and smooth leaves whose breakdown was hypothesized to be comparable with that of *Eucalyptus*. Syzygium has also been used in decomposition studies in the region and was chosen for comparison (Mathooko et al., 2000).

Abscised leaves were air dried at room temperature for two weeks to attain constant mass before weighing. Thereafter leaves were enclosed in either coarse-mesh (10 mm) or fine-mesh (0.5 mm) bags (~4 g of each species per bag). Unlike the coarse-mesh, the fine-mesh was meant to exclude shredders so that breakdown rates could be attributed to microbial processing only (Gessner and Chauvet 2002). Before deployment 252 litterbags were arranged into sets of four replicates per litterbag type (2 types) per plant species (3 plant species) in each stream (10 streams = 240 litterbags) per season. The 12 extra litterbags were used to determine initial ash-free-dry mass (AFDM) for each leaf species. Litterbags were deployed at each site and secured by nylon lines sufficiently far apart to avoid overlap. Bags were retrieved after 56 days. Loss of invertebrates and leaf fragments was avoided by enclosing bags in 300-μm mesh net before removal. Individual bags were preserved in 75% ethanol.

2.4 Invertebrates

Stream invertebrate kick samples were collected once each during the dry (January-February 2011) and wet season (May-July 2011). Sample sites included ten least-disturbed forest (FOR) streams located in Transmara Forest, seven agriculture (AGR) streams in catchments draining agriculture lands and seven mixed (MIX) streams in catchments that were partly forest and partly agriculture (Figure 4.1). At each site and occasion, and along a 100-m reach of stream, five random kick samples each were collected, respectively, from riffles and pools (total 10 samples) using a dip net of 300 μm mesh-size. An area covering approximately 30 × 50 cm was disturbed rapidly for 10 seconds and all contents from the net preserved in 75% ethanol and stored in polythene bags. The rapid sampling method was necessary to capture large invertebrates, especially fast moving crabs (Magana et al., 2012).

2.4.1 Community structure and functional composition

Kick samples: After sorting from debris, invertebrates were identified to the lowest possible taxonomic level or morphospecies with a series of guides (Day et al. 2002a,b,c; de Moor et al., 2003a,b; Stals and de Moor, 2007; Merritt et al., 2008). All samples from pools and riffles were composited for each site per season. Invertebrates were assigned into four FFGs – collectors, predators, scrapers and shredders according to Masese et al. (2014a). In summary, gut contents were removed under 40 × magnification onto a glass slide before being mounted with polyvinyl lactophenol. Using a compound microscope, estimates were made of percent of different food items in guts, assumed to be 100% full. Food items identified included vascular plant material (VPM - particles >1 mm), coarse particulate organic matter (CPOM – particles from 50 μm to 1 mm), fine particulate organic matter (FPOM – particles <50 μm), algae and animal material. Shredders were invertebrates whose gut contents comprised mainly leaf and wood fragments >1 mm, while predators fed mainly on animal material. Collectors were assumed to consume CPOM and FPOM and scrapers FPOM and algae. For invertebrates whose guts were empty or food items indistinguishable, literature was used to determine FFG (Dobson et al., 2002, Merritt et al., 2008).

Litterbag samples: During retrieval, some litterbags were lost to sedimentation and this reduced the number of analyzed replicates to 3 per mesh size per leaf species. Invertebrates in litterbags were sorted under a stereo microscope and identified to the lowest possible taxonomic level. Leaves and leaf fragments from bags were rinsed and oven-dried at 105°C for 24-48 h to yield dry mass and then ashed at 550 °C for 4 h and reweighed to calculate percent ash and AFDM.

2.5 Statistical analyses

Habitat conditions expressed as percentages were arcsin [$\sqrt{(x/100)}$] transformed while physico-chemical variables, except pH, were ln (x+1) transformed before analysis to meet assumptions for parametric tests (Zar, 1999). Invertebrate count data were ln (x+1) transformed for parametric tests while un-transformed data were used for non-parametric tests. The arcsin ($\sqrt{(x/100)}$) transformation is appropriate for data expressed as a percentage while $\ln(x+1)$ transformation is appropriate for count data with values less than 10 (Zar, 1999). All statistical analyses were performed with Statistica (Version 7, StatSoft, Tulsa, Oklahoma), unless otherwise indicated. Two-way ANOVA was used to test for differences in physico-chemical variables, riparian and organic matter variables among land uses (AGR, MIX, FOR) and seasons (dry and wet) with land use and seasons as main factors and a land use×season interaction term. Where there were no significant seasonal differences, data were pooled and one-way ANOVA tested for differences among land uses followed by Tukey multiple *post hoc* comparisons of the means. Principal Component Analysis (PCA) was used to summarize variation in physico-chemical variables, riparian and organic matter characteristics among land uses and sites.

Community structure and functional composition of invertebrates in kick samples and coarse-mesh litterbags were described as abundance (number of individuals per sample or litterbag), taxa richness (number of taxa per sample or litterbag) and the four FFGs -collectors, predators, scrapers and shredders. General linear models (GLMs) were used to test differences in total abundance of all taxa, richness of all taxa (number of taxa), shredder and non-shredder richness and shredder abundance in kick samples between seasons (dry and wet) and among land uses, followed by *post hoc* Tukey tests to identify differences among land uses and land use×season interactions.

Non-metric multidimensional scaling (NMDS) was used to visualise invertebrate community structure and functional composition in the kick samples among land uses and seasons. Dissimilarity matrices based on the Bray–Curtis coefficients (Bray and Curtis, 1957) were derived for 4 data sets: taxa presence–absence data of taxa richness for all invertebrates, un-transformed abundances for all invertebrate taxa, un-transformed abundances for the four functional groups, and taxa presence–absence for the four functional groups. Goodness of fit of the ordination was assessed by the magnitude of the associated stress value; a value of <0.2 corresponds to a good ordination (Kashian et al., 2007). The percentage contribution of each taxon to the overall dissimilarity between agriculture and forest land use were quantified by the similarity percentages (SIMPER) routine in Paleontological Statistics (PAST) software package (Version 2.17; Hammer et al. 2001). SIMPER is a strictly pairwise analysis between two factor levels (Clarke and Warwick, 2001); in this case agriculture and forest.

Leaf breakdown rates were estimated using an exponential decay model $Wt = W_0 e^{-kt}$ (Wt = remaining AFDM at time t (56 days); W_0 = initial AFDM; $-k$ = decay rate (Boulton and Boon, 1991). Breakdown rates for fine- (kf) and coarse-mesh (kc) litterbags were calculated separately. To determine the effect of excluding potential invertebrate shredders from fine-mesh litterbags on breakdown, kc /kf coefficients were calculated for each stream. Similarly, ki /kr coefficients (i for impacted [agriculture] and r for reference [forest] streams, respectively) were calculated for coarse- and fine-mesh litterbags (Gessner and Chauvet, 2002) to determine the effect of land use on shredder and microbial breakdown, respectively. GLMs were used to compare total taxa richness, shredder and non-shredder taxa richness, and shredder abundance in coarse-mesh litterbags for forest and agriculture streams. Four-way ANOVA explored variation in leaf breakdown rates (-k) with season (wet and dry), land use (forest and agriculture), leaf species (*Croton*, *Syzygium* and *Eucalyptus*) and treatment by mesh size (fine- or coarse-mesh litterbags) as the main factors, including interactions. Because of lack of season×land use×leaf species×treatment-by-mesh-size interactions and to partition seasonal influences on breakdown, three-way ANOVAs were re-run separately for the wet and dry seasons with land use, leaf species and treatment by mesh size as the main factors, including interactions.

3. Results

3.1 Environmental conditions

Streams in agriculture and forest catchments showed differences in physico-chemical and organic matter characteristics, with mixed streams being intermediate (Table 4.2). Factor 1 in the PCA ordinations accounted for most variation, distinguishing most forest from agriculture sites (Figure 4.2abcd). Variables most related to Factor 1 were related to water quality (temperature, turbidity TSS) and stream size (discharge, width and depth), which increased towards agriculture and mixed streams. Riparian and organic matter characteristics (% canopy cover, % leaf litter and litter biomass) increased towards forest streams (Figure 4.2abcd). Canopy cover was >80% in most forest streams with decomposing leaf litter and woody debris dominating in pools. In agriculture streams, canopy cover was less than 50% and discontinuous in stream reaches frequented by people and livestock. Area covered by decomposing litter (% leaf litter) did not differ between seasons but was higher in forest than in both agriculture and mixed streams (one-way ANOVA, $F_2 = 28.62$, $p < 0.001$). There were significant differences in POM standing stocks among the land uses (one-way ANOVA, $F_2 = 8.44$, $p < 0.05$), but not with season (Table 4.2).

Factor 2 was associated with nutrients (SRP and TDN) and DOC, which increased in agriculture streams. Forest streams were cooler (mean temperature: 14.4±0.58 °C) than agriculture streams (mean: 18.7±2.88 °C). TSS, turbidity, TDN, DOC, Cl^-, SO_4^{-2}, HCO_3^-, DSi and NH_4-N had significantly ($p < 0.05$) higher values in agriculture streams than in forest streams (Table 4.2). TSS, turbidity, TDN, NH_4-N and DOC were consistently higher during the wet season while Cl^-, SO_4^{-2}, and HCO_3^- were higher during the dry season across the three land uses. Fe^{++}, Na^+, K^+ and PO_4-P, NO_3-N and conductivity showed no seasonal variation but significantly ($p < 0.05$) higher values in agriculture streams compared with forest.

Table 4.2. Mean (±SE) values for physico-chemical variables, riparian conditions and organic matter for streams within the three land use categories; agriculture, mixed and forest. Means for physico-chemical variables that displayed significant differences (Two-way ANOVA) between the dry and wet seasons have been presented separately. Similar superscripts on means indicate no significant difference for variables and organic matter among land use types - *post hoc* Tukey tests; agriculture - n = 7, mixed - n = 7, forest - n = 10 per season. *designates significant differences at $p < 0.05$.

Physico-chemical Variables		Land Use			F	p – value
		Agriculture	Mixed	Forest		
Organic matter fractions						
Leaves (g/m^2)		100. 1±14.8^a	117.5±14.0^b	162.9±15.2^b	5.02	0.012*
Sticks/wood (g/m^2)		55.8±7.4^a	62.1±11.7^a	60.22±7.4^a	0.14	0.870
Fruits and flowers (g/m^2)		15.9±5.5^a	17.5±4.8^a	24.2±5.2^b	7.20	0.028*
Seeds (g/m^2)		7.4±0.8^a	3.4±0.2^a	5.3±0.9^a	0.45	0.799
Standing stock (g/m^2)		145.6±13.6^a	169.1±19.2^b	233.0±14.4^b	8.44	0.040*
% Litter		36.8±4.5^a	45.2±3.3^a	78.1±4.4^b	28.62	<0.001*
% Canopy cover		45.4±2.6^a	54.2±2.9^a	80.7±4.5^b	27.96	<0.001*
No seasonal variation						
Ca (mg/l)		5.2±0.1^a	4.3±0.1^a	2. 7±0.5^a	5.62	0.051
Mg (mg/l)		1.3±0.1^a	1.4±0.1^a	0.9±0.2^a	4.10	0.128
K (mg/L)		9.2±0.8^a	4.6±0.3^b	4.3±0.2^b	8.31	0.016*
Na (mg/L)		11.8±1.3^a	7.7±0. 5^b	6.2±0.5^b	16.00	<0.001*
Fe (mg/l)		2.0±0.9^a	2.4±1.2^a	0.9±0.4^b	8.57	0.015*
NO$_3$-N (mg/l)		1.2±0.3^a	1.2±0.4^a	0.3±0.1^b	11.62	0.003*
PO$_4$-P (mg/l)		0.2±0.1^a	0.2±0.2^a	0.1±0.01^b	8.97	0.011*
pH		5.0±0.8^a	6.5±1.1^a	6.7±0.4^a	0.54	0.764
DO (mg/l)		6.7±0.7^a	5.7±1.0^a	7.3±0.2^a	3.42	0.181
Conductivity (μS/cm)		167.3±29.6^a	116.1±25.8^{a,b}	87.3±11.2^b	10.021	0.007*
Temperature (°C)		18.7±2.9^a	16.4±1.9^{ab}	14.4±0.6^b	7.61	0.022*
TP (mg/l)		0.3±0.1^a	0.3±0.1^a	0.2±0.1^a	2.94	0.231
TN (mg/l)		1.3±0.25^a	1.6±0.34^a	0.6±0.1^b	9.61	0.008*
SRP (μg/l)		17.2±7.6^a	18.7±6.5^a	1.3±0.9^b	9.61	0.008*
Seasonal variation						
TSS (mg/l)	Dry	85.7±24.6^a	73.3±8.0^a	22.7±7.6^b	26.17	<0.001*
	Wet	254. 9±74.1^a	173.5±14.6^a	36.6±20.5^b	20.21	<0.001*
Turbidity (NTUs)	Dry	64.5±9.5^a	38.7±6.1^a	7.8±0.8^b	8.89	0.012*
	Wet	122.8±23.3^a	77.1±9.2^a	21.1±1.2^b	7.68	0.023*
TDN (mg/l)	Dry	1.1±0.3^a	1.2±0.3^a	0.3±0.2^b	6.78	0.041*
	Wet	1.3±0.4^a	2.8±0.5^{ab}	1.0±0.4^b	10.73	0.005*
POM (mg/l)	Dry	25.0±13.5^a	38.5±14.6^a	10.6±8.5^a	0.089	0.956
	Wet	70.7±17.1^a	84.6±40.1^{ab}	21.9±10.2^b	6.45	0.045*
DOC (mg/l)	Dry	3.6±0.9^a	3.9±0.3^a	2.7±0.4^b	8.07	0.016*
	Wet	8.1±0.92^a	4.2±1.23^a	3.5±0.6^b	14.11	0.001*
Cl (mg/l)	Dry	5.6±1.4^a	6.1±1.2^{ab}	3.8±0.4^b	6.38	0.046*
	Wet	3.9±0.6^a	5.1±1.2^a	1.0±0.8^b	6.72	0.035*
SO$_4$ (mg/l)	Dry	3.8±0.6^a	6.1±1.8^b	3.2±0.7^a	6.78	0.041*
	Wet	2.7±0.5^a	4.0±2.1^a	0.5±0.4^b	7.78	0.020*
HCO$_3$ (mg/l)	Dry	59.3±18.7^a	56.7±6.3^{ab}	25.0±5.6^b	10.73	0.00
	Wet	23.3±6.9^a	29.5±2.1^a	13.1±10.8^a	3.02	0.223
DSi (mg/l)	Dry	25.9±2.4^a	39.2±2.8^a	27.7±2.7^b	6.8	0.040*
	Wet	38.5±16.4^a	25.1±9.3^a	9.0±6.6^b	4.88	0.039*
NH$_4$-N (μg/l)	Dry	28.2±13.6	51.3±32.4^a	11.2±1.4^b	10.73	0.005*
	Wet	11.6±7.7^a	14.2±8.6^a	2.5±1.2^b	7.61	0.022*
% Riffle	Wet	49.0±2.3^a	48.6±5.3^a	53.3±5.4^a	0.36	0.702
	Dry	53.6±4.4^a	54.0±5.8^a	43.6±2.0^a	1.02	0.325

76

| % Pool | Wet | 51.2±4.3[a] | 47.9±3.4[a] | 46.7±4.5[a] | 0.40 | 0.678 |

Table 4.2 Continued

Physical variables		Agriculture	Mixed	Forest	F	p – value
	Dry	52.0±4.8[a]	51.8±2.3[a]	62.9±5.7[a]	2.03	0.118
Depth (m)	Wet	0.2±0.1[a]	0.2±0.1[a]	0.3±0.1[a]	0.10	0.906
	Dry	0.2±0.1[a]	0.1±0.02[a]	0.1±0.02[a]	2.20	0.137
Width (m)	Wet	2.8±0.4[a]	2.3±0.3[a]	2.1±0.3[a]	1.12	0.346
	Dry	3.6±0.9[a]	3.1±1.1[a]	2.4±0.9[a]	0.44	0.651
Discharge (m^{3}s^{-1})	Wet	0.6±0.1[a]	0.5±0.4[a]	0.4±0.3[a]	1.56	0.233
	Dry	0.2±0.1[a]	0.1±0.1[a]	0.1±0.1[a]	2.85	0.081

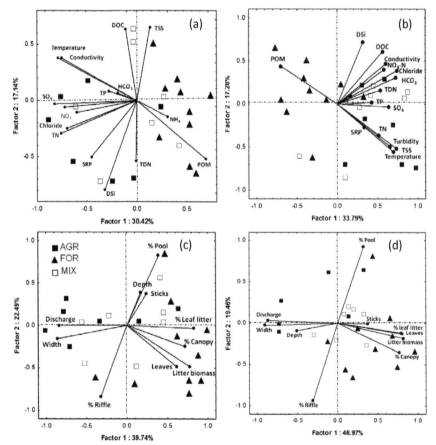

Figure 4.2. Principal component analysis on physico-chemical variables (a, b) and riparian and organic matter characteristics (c, d) during the wet (a, c) and dry (b, d) seasons. Variation explained by each axis (r^2, expressed as a percentage) is shown; FOR- n = 10, AGR - n = 7, MIX - n = 7 per season.

3.2 Community structure

A total of 25,887 individuals belonging to 109 taxa were collected. Of the 109 taxa, 81and 93 were collected during the dry and wet seasons, respectively. Total abundance was higher during the wet than dry season ($F_{1,2} = 7.03$, $p < 0.05$) but did not differ among land uses. Taxa richness was highest in forest streams (richness, $F_{1,2} = 3.02$, $p < 0.05$), but did not differ between seasons (Figure 4.3). Ninety-six taxa were found in forest streams, 70 in mixed streams and 60 in agriculture streams. Fifty-four taxa occurred in all land uses, 21 taxa were restricted only to forest streams, six to mixed streams and two to agriculture streams. NMDS ordinations of abundance data indicated separation of most FOR from AGR sites (Figure 4.4a,b), and clear separation of FOR and AGR sites for presence–absence of taxa (Figure 4.4c,d) indicating that the difference in invertebrate communities was not only as a result of differences in relative abundance, but was also caused by differences in taxa richness. SIMPER identified Simuliidae, *Tricorythus tinctus* and *Pseudocloeon* sp. as taxa most influencing the difference between FOR and AGR communities during the wet season (Table 4.3). Other important taxa included Chironominae, *Afronurus* sp., *Cheumatopsyche thomassetti*, *Tubifex* sp., *Hydropsyche* sp. and *Afrocaenis* sp. The same taxa were important during the dry season, except for a decrease in abundance of most taxa in both FOR and AGR streams.

3.3 Functional composition

A total of 19 shredders, 26 collectors, 21 scrapers and 43 predator taxa were identified in the study area. All 19 shredder taxa occurred in forest streams, while none were restricted to the streams in mixed or agriculture catchments. Eight shredder taxa were restricted to forest streams and another eight widespread across all land uses, with a seasonal×land use interaction ($F_{1,1} = 3.94$, $p < 0.05$). In both seasons total number of taxa and number of shredder taxa were higher in forest than in agriculture streams. Non-shredder richness did not vary among land uses during the dry and wet seasons (Figure 4.3). Shredder abundance was higher in forest compared with agriculture and mixed streams only during the wet season (Figure 4.3). *Potamonautes* sp.1 was the most abundant shredder in forest streams with a mean abundance per sample of (29.6±5.0), followed by *Lepidostoma* sp. (14.6±2.4) and *Acanthiops* sp. (8.3±3.4). In agriculture streams, Pyralidae sp. 1 dominated abundance (9.3±3.6) followed by *Lepidostoma* sp. (5.9±1.6), then *Tipula* sp.1 (4.6±1.3).

NMDS ordination plots for un-transformed abundances and un-transformed presence-absence (taxa richness) data for the four functional groups were similar during the wet and dry seasons and separated most forest streams from agriculture streams on both axes (Figure 4.5abcd). While the responses among the various FFGs were variable, abundance and taxa richness of each FFG responded similarly to land use influences.

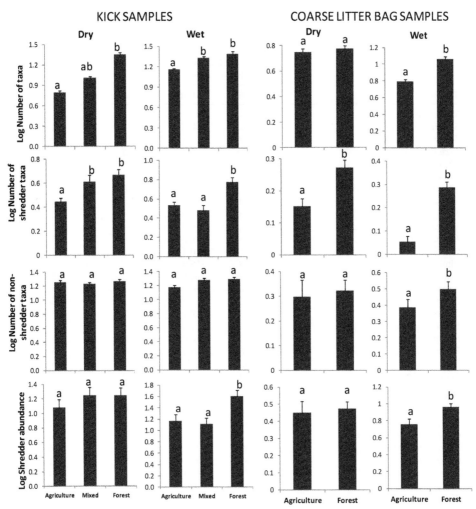

Figure 4.3. Mean (± S E) number of taxa, number of shredder taxa, number of non-shredder taxa and shredder abundance (transformed to the log[x+1]) during the dry and wet seasons at agriculture, mixed and forest sites for kick samples and at agriculture and forest sites for colonized coarse-mesh bags (different letter indicates significant differences among land uses). Note the differing range in y-axes. For kick samples FOR- n = 10, AGR - n = 7, MIX - n = 7 per season; for coarse litterbag samples, n = 15 for all sites per season.

Table 4.3. Top-ranked SIMPER contributors to % dissimilarity in the composition of invertebrate taxa between forest (FOR) and agriculture (AGR) sites during the wet and dry seasons. Mean abundance was calculated as the number of individuals/ m² for all sites per land use. FOR- n = 10, AGR - n = 7 per season.

Taxon	Mean Abundance (individuals / m²		% Contribution	% Cumulative
	FOR	AGR		
Wet season				
Simuliidae	412.0	148.0	17.5	17.5
Tricorythus tinctus	2.8	115.0	7.7	25.2
Pseudocloeon sp.	65.5	106.0	6.5	31.7
Chironominae	65.8	63.6	4.5	36.1
Afronurus sp.	11.3	78.4	4.4	40.5
Cheumatopsyche thomassetti	9.7	50.9	3.0	43.5
Tubifex sp.	5.4	30.6	2.5	46.0
Hydropsyche sp.	10.6	34.6	2.4	48.4
Afrocaenis sp.	36.3	10.0	2.4	50.8
Pisidium sp.	33.4	6.6	2.4	53.2
Orthocladinae	23.7	20.4	2.1	55.3
Elminae	6.8	15.0	1.3	56.6
Acanthiops sp.	14.6	9.9	1.2	57.8
Lestidae	0.0	9.4	1.2	59.0
Lepidostoma sp.	9.6	15.0	1.1	60.2
Dicercomyzon sp.	0.8	22.4	1.1	61.3
Cheumatopsyche sp. 1	10.4	12.6	1.1	62.3
Potamonautes sp.1	15.8	0.9	1.0	63.3
Wormalidia sp.	4.6	10.6	0.9	64.3
Neoperla spio	13.5	1.1	0.9	65.2
Tanypodinae	14.6	9.3	0.9	66.1
Dry season				
Simuliidae	240.5	83.3	8.0	8.0
Pseudocloeon sp.	46.5	83.5	7.8	15.8
Cheumatopsyche thomassetti	8.2	80.8	7.2	23.0
Tricorythus tinctus	4.0	98.0	6.9	29.9
Chironominae	44.0	63.5	5.9	35.9
Hydropsyche sp.	12.8	45.3	5.1	41.0
Tanypodinae	14.1	32.5	3.4	44.3
Afronurus sp.	4.5	33.0	3.1	47.4
Pisidium sp.	28.8	5.5	2.8	50.2
Caenis sp.	2.4	15.3	2.6	52.8
Cheumatopsyche sp.	19.0	8.8	1.8	54.6
Scirtidae	16.0	4.4	1.6	56.2
Afrocaenis sp.	10.8	5.8	1.3	57.5
Orthocladinae	8.3	8.8	1.3	58.8
Neoperla spio	12.6	1.8	1.1	59.9
Wormalidia sp.	10.4	4.5	1.1	61.0
Elminae	8.3	15.1	1.1	62.1
Propistomatidae	6.1	5.4	0.9	63.0
Dicercomyzon sp.	0.0	8.0	0.9	63.9
Lepidostoma sp.	9.5	4.9	0.9	64.7
Pyralidae sp.1	0.9	7.3	0.7	65.5
Acanthiops sp.	0.0	10.8	0.7	66.1
Potamonautes sp 1	6.9	0.8	0.6	66.7
Polycentropus sp.	5.3	0.6	0.5	67.2

3.4 Leaf litter breakdown

Across seasons and land use, breakdown rates were species-specific and generally displayed similar trends with *Croton* having the fastest decay rate followed by *Syzygium* then *Eucalyptus* (Figure 4.6). Breakdown rates (-k) were higher in coarse- than in fine-mesh litterbags for the three species and in the two land uses, except for *Eucalyptus* during the wet season. Plant species and mesh-size influenced breakdown rates during the wet and dry seasons, but land use was only important during the wet season (Table 4.4). However, lack of interactions between and among the three factors implies that shredder and microbial contributions to litter breakdown were not consistent for all plant species across sites in the two land uses. During the wet season, there interaction between land use and treatment by mesh size was significant while that between land use and leaf species was marginally significant $(0.05 > p < 0.1)$. This indicates that response of breakdown, by both shredders and microbes, to land use was dependent on leaf quality. For instance, *Croton* displayed the greatest response to microbial processing in agriculture streams during the dry and wet seasons. However, shredder processing were similar for *Croton* and *Syzygium* during both seasons and in the two land uses, except in agriculture streams during the wet season (Figure 4.6). Total number of taxa, number of shredder taxa and shredder abundance in coarse-mesh litterbags were higher in forest than agriculture streams during the wet season. During the dry season, land use effects were not evident; shredder abundance was not affected by land use, although shredder taxa richness was higher in forest streams.

The relative contribution of shredders to breakdown was higher in forest streams for the three leaf species and was highest for *Croton*, followed by *Syzygium* and then *Eucalyptus*, except in agriculture streams when the contribution was highest in *Syzygium* (Table 4.5). In coarse-mesh litterbags during the dry season the ki/ kr coefficients showed that shredder breakdown of leaf litter was limited in agriculture streams for *Croton* (ki /kr=0.84) but not for *Eucalyptus* (ki /kr=1.39) or *Syzygium* (ki/ kr=0.99). During the wet season the effects on shredder breakdown were more evident for *Croton* (ki /kr=0.82), and *Syzygium* (ki/ kr = 0.43) (Figure 4.6). Conversely, during the dry season microbial breakdown was enhanced in agriculture streams for *Croton* (ki /kr = 1.26) and *Eucalyptus* (ki /kr = 1.38). During the wet season microbial processing in forest and agriculture streams were similar for *Croton* (ki /kr = 1.08) and *Eucalyptus* (ki /kr=1.01) and reduced in agriculture streams for *Syzygium* (ki /kr = 0.60).

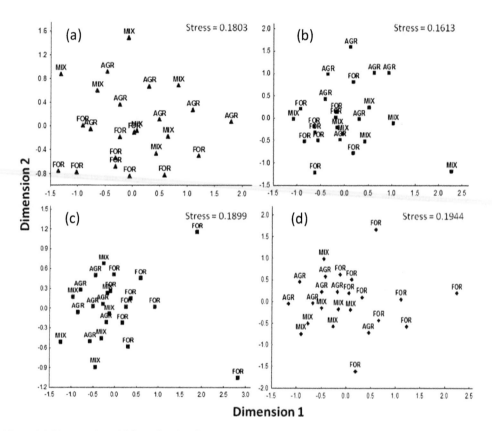

Figure 4.4. Non-metric multidimensional scaling (NMDS) plots of invertebrate community composition based on abundance data of the various taxa during (a) wet season and (b) dry season, and species presence-absence data of various taxa during (c) wet season and (d) dry season in the three land uses: forest (FOR), agriculture (AGR) and mixed (MIX). Presence-absence data responded more strongly to land use compared with abundance data.

82

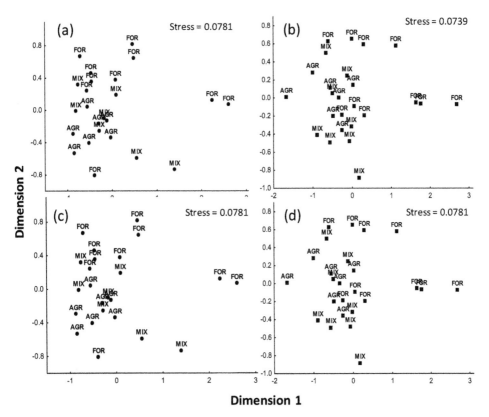

Figure 4.5. Non-metric multidimensional scaling (NMDS) plots of invertebrate composition of the four functional groups based on abundance data during (a) wet season and (b) dry season, and based on presence-absence data during (c) wet season and (d) dry season in the three land uses: forest (FOR), agriculture (AGR) and mixed (MIX). Abundance and presence-absence of the functional feeding groups responded similarly to land use during the two seasons.

Table 4.4. Results of three-way ANOVA exploring variation in leaf breakdown rates (-k) with land use (forest and agriculture), leaf species (*Croton*, *Syzygium* and *Eucalyptus*) and treatment by mesh size (coarse and fine mesh) during the wet and dry season. Degrees of freedom (df), sums of squares (SS), F-statistic and *p*-values are shown (significant values at $p < 0.05$ level in bold).

Source of variation	df	SS	F	*p*
Wet season				
Land use (LU)	1	1.53×10^{-3}	6.35	**0.013**
Leaf species (LS)	2	7.83×10^{-3}	16.25	**<0.001**
Treatment by mesh size (Tr)	1	4.08×10^{-3}	16.93	**<0.001**
LU × LS	2	1.25×10^{-3}	2.59	0.079
LU × Tr	1	1.09×10^{-3}	4.53	**0.035**
LS × Tr	2	1.73×10^{-4}	0.36	0.699
LU × LS × Tr	2	3.00×10^{-4}	0.63	0.538
Error	168	4.05×10^{-2}		
Dry season				
Land use (LU)	1	3.88×10^{-6}	0.03	0.875
Leaf species (LS)	2	3.40×10^{-3}	10.95	**<0.001**
Treatment by mesh size (Tr)	1	1.01×10^{-2}	64.88	**<0.001**
LU × LS	2	2.80×10^{-4}	0.90	0.408
LU × Tr	1	1.34×10^{-4}	0.86	0.354
LS × Tr	2	4.11×10^{-3}	1.32	0.269
LU × LS × Tr	2	1.12×10^{-3}	0.36	0.699
Error	168	2.61×10^{-2}		

Table 4.5. Relationship between leaf litter breakdown rates between coarse- (kc) and fine-mesh (kf) sizes (kc /kf coefficient) across all streams used for litterbag experiments. The coefficients show the contributions of shredders (coarse-mesh) relative to microbes (fine-mesh) to litter breakdown. Coefficients have been provided for each leaf species separately for the dry and wet seasons. Means are for each leaf species per land use and season; n = 3 for site means per leaf species per season. The site numbers correspond to litterbag experiments site numbers in Figure 4.1.

Land use/ Site	Dry season			Wet season		
	Croton	*Eucalyptus*	*Syzygium*	*Croton*	*Eucalyptus*	*Syzygium*
Forest						
F1	1.89	1.56	1.23	2.95	0.98	1.91
F2	1.22	0.82	1.47	1.95	1.54	2.60
F3	1.73	1.53	1.11	1.21	0.90	1.28
F4	1.80	1.55	1.66	1.60	0.59	1.56
F5	1.63	1.94	2.15	2.98	1.67	2.50
Means	**1.65**	**1.48**	**1.52**	**2.14**	**1.14**	**1.97**
Agriculture						
A1	0.64	0.85	1.10	0.76	0.31	1.35
A2	1.28	1.44	2.38	1.45	1.39	2.46
A3	1.52	0.84	2.03	1.75	0.88	1.96
A4	1.34	1.16	0.61	1.88	0.26	2.07
A5	1.74	1.51	2.12	1.53	0.95	0.66
Means	**1.30**	**1.16**	**1.64**	**1.48**	**0.76**	**1.70**

4. Discussion

Assessing the influence of catchment land use, riparian and in-stream disturbances (livestock activity, bathing and laundry washing) on ecological condition while accounting for natural and seasonal variations is essential for natural resource management. In this study, human-induced changes in physico-chemical variables, nutrient concentrations and organic matter (leaf litter) were propagated to consumers. This skewed the contribution of shredders to leaf litter breakdown in agriculture streams and that of microbes in forest streams.

There were increased concentrations of major ions, turbidity, TSS, conductivity, temperature and dissolved nitrogen in streams in agriculture landscapes compared with those in forest. Low temperature in forest streams was due to high canopy cover (above 80%) provided by natural riparian vegetation, which protected the streams from direct insolation. Most native riparian trees tend to grow over the stream whereas Eucalyptus spp. are more columnar to pyramidal in shape.

In streams that drain catchments of similar geology, variability in electrical conductivity is indicative of anthropogenic activities. Whereas turbidity, TSS and bioavailable nitrogen showed a relationship with catchment land use, they have been found to be more responsive to local human and animal activities in the study area (Minaya et al., 2013; Kilonzo et al., 2013), implying that even among agriculture streams variability is expected.

Seasonal influences on run-off and discharge emphasized differences among land uses in some of the variables such as nitrogen, turbidity and suspended sediments. Increased in-stream activities by livestock and people observed in streams in agriculture landscapes were major sources of sediments during the dry season (see Mathooko, 2001; Yillia et al., 2008), whereas unpaved roads and footpaths became major source of sediments during the wet season. Since human activities were irregular along agriculture streams, physico-chemical conditions were patchy resulting in greater inter-site differences during the dry season.

Reduced canopy cover due to removal of indigenous riparian vegetation and its replacement by *Eucalyptus* spp. in agriculture streams reduced the quality of litter. The allelopathy of *Eucalyptus* spp. reduce diversity and alter structural attributes of native vegetation (May and Ash, 1990). This could further reduce the quality of leaf litter in agriculture streams. Compared with particulate fractions, dissolved organic matter was higher in agriculture streams, likely contributed by enhanced primary production and inputs of human and animal wastes.

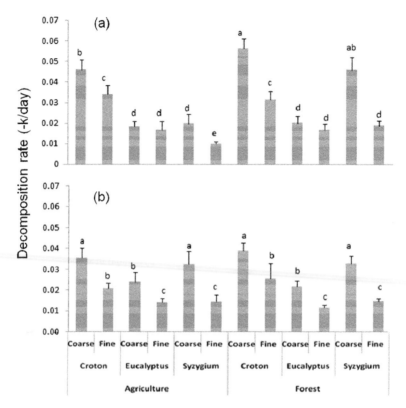

Figure 4.6. Mean (± S E) showing the effects of land use, leaf species and treatment by mesh size on leaf decomposition rate (-kday[-1]) for the (a) wet and (b) dry seasons. Similar trends are displayed for decomposition of the three leaf-types across land uses during the wet and dry seasons with *Croton* decomposing fastest followed by *Syzygium* and then *Eucalyptus*; n = 15 per leaf species, per land use, per season.

4.1 Invertebrate assemblages

Numerical abundance of invertebrates was a poor measure of disturbance and the abundance of most taxa was considerably lower during the dry season. However, these results should be interpreted with caution given the short period of the study. Although natural conditions influence taxon richness and abundance, it is likely that human activities exacerbated the effects observed in this study. Land use activities relating to road building, agriculture and settlements have been found to influence the quantity of runoff and sediments that enter recipient water bodies during the rains (Wang and Lyons, 2003; Donohue and Irvine, 2004). However, during the peak of the dry season, conditions can worsen in streams and rivers because, as discharge declines fines accumulate, temperatures rise and dissolved oxygen becomes limiting (Harrison and Hynes, 1988; Mathooko and Mavuti, 1992; Shivoga, 2001). Long-term studies are needed in these streams to elucidate temporal trends and to differentiate between natural variability arising from insect adult emergence and human-induced declines in water and habitat quality.

Taxa richness varied with land use and season with higher numbers in forest streams compared with agriculture streams. However, there was a decrease in land use related influences during the wet season and an increase in reach-scale influences during the dry season. During the wet season, spates and scouring make the stream substratum more uniform and reduces inter-site differences in community structure (Leung et al., 2012). Flow increases connectivity enabling taxa to colonize new areas. There is also a notable reduction in human and livestock disturbance during the wet season when harvested rainwater is the main source of water for domestic and livestock use, reducing the pressure on streams (pers. obser.). In contrast, human and livestock activities (bathing, laundry washing and livestock watering) in streams are higher during the dry season (Yillia et al., 2008). As this happens at select reaches, impacts are irregularly spaced, resulting in greater spatial variability and among land use differences. Rheophilic taxa (e.g., Simuliidae, *Tricorythus tinctus*, *Cheumatopsyche* spp.) responded to land use influences, signifying the important role played by flow in structuring invertebrates in these streams. As flow remains interstitial and exposes riffles, rheophilic taxa are disadvantaged while burrowing and pool taxa such as Tubificidae and Chironomidae thrive (Mathooko et al., 2005; Masese et al., 2009a). Studies that explore flow-ecology relationships in these highly hydrologically variable systems are needed to inform sustainable water resources development (McClain et al., 2014).

During the dry and wet seasons, abundance and taxa richness of FFGs responded similarly to land use influences. Shredders exhibited highest diversity in forest streams while collectors were dominant in agriculture streams. Changes in water quality and organic matter characteristics can explain the skewed distribution of shredder taxa in forest streams. Although shredder numbers are quite variable in the tropics, water temperature and leaf litter characteristics play important roles (Yule et al., 2009; Boyero et al., 2011a,b,c). In this study, agriculture streams were warmer and with higher nutrient concentrations and suspended sediments, and leaf litter was mainly of the exotic *Eucalyptus* species.

4.2 Litter breakdown and ecosystem functioning

The importance of microbes relative to shredders as agents of breakdown in agriculture streams compared with forest streams connects land use with stream processes, even though microbial breakdown was not measured directly but inferred from the exclusion of shredders in fine-mesh litterbags. Shredders, although important in agriculture streams during the dry season, were generally disadvantaged for *Eucalyptus*. Nevertheless, the combined effects of increased microbial activity and high abundance and widespread distribution of certain shredders such as *Tipula* sp.1 and *Lepidostoma* sp. are interpreted to have contributed to increased breakdown rates in some agriculture streams during the dry season. This also implies that both seasonal and reach-scale influences were highly relevant for invertebrate assemblage structure, and consequently litter decomposition as a metric of function. Microbial processing responded more to higher ambient nutrient concentrations in agriculture streams for the higher quality *Croton* than for *Syzygium* and *Eucalyptus*. These results conform to the hypothesis that leaf quality mediates the effects of elevated nutrients on microbial processing (Ardón et al., 2006). The poorer quality (C:N ratio, 63.7-89.7) of *Eucalyptus* for shredders compared with *Croton* (10.6-16.2) and *Syzygium* (38.9-50.9) (unpublished results) is supported by this study. In most sites, the kc/kf coefficients were highest in *Croton*, indicating that invertebrate –mediated breakdown was also affected by litter quality.

Even though reported for individual leaf species, the kc/kf coefficients for *Croton* and *Syzygium* incubated in forest streams are comparable with a value of 1.98 reported for a litter mixture in a forest stream in Borneo (Yule et al., 2009). The decomposition rates of *Syzygium* in coarse-mesh litterbags found in my agriculture streams during the wet season (0.010-0.043 d^{-1}) are also comparable to those found by Dobson et al. (2003, 0.022 d^{-1}). Like the Bornean forest stream (Yule et al., 2009), forest streams in this study contained diverse and abundant shredder taxa.

The results of this study have important implications on the management of riverine ecosystems in the region. First, catchment-scale pressures influence ecosystem functioning as can be inferred from the restricted occurrence of 21 taxa, including eight shredder taxa, in forest streams. The importance of reach-scale influences was illustrated by inter-site differences in some physico-chemical variables and assemblage characteristics among streams with similar catchment land uses. To maintain ecological integrity of these streams management actions addressing both catchment- and reach-scale are required. Second, shredder diversity response to changes in allochthonous POM quality demonstrates the potential to affect nutrient cycling when indigenous vegetation is replaced by *Eucalyptus* species. Accumulation of significant amounts of slowly decomposing leaf litter reduces the capacity to support diverse consumers and complex food webs via detrital pathways. Replacing riparian forests with exotic tree species also increases available light, stimulating benthic algal production and overall ecology of shallow streams and a shift in the relative importance of allochthonous relative to autochthonous sources of carbon for food webs. Thirdly, through its controls on discharge, depth and material load (particulate and dissolved) seasonality accentuate differences in environmental conditions among land uses and within sites among agriculture and mixed land uses.

Conclusions

In addition to highlighting the applicability of leaf litter processing and the composition of invertebrate FFGs as functional and structural indicators, respectively, of ecological health, this study highlights interactions among catchment land use, riparian activities and seasonality as drivers of ecosystem functioning in upland tropical streams. While catchment land use is an important determinant of temperature and litter biomass in the studied streams, reach-scale influences that affect leaf litter quality through exotic introductions and reduced water quality were equally important in structuring invertebrate communities, with effects propagating to consumers and processing of leaf litter. The relative differences in breakdown rates among the three plant species remained unaltered in both agriculture and forest streams irrespective of mesh-size; *Croton* was fastest followed by *Syzygium* while *Eucalyptus* was the slowest. The fast leaf breakdown in fine-mesh litterbags observed in this study has been observed in other tropical stream studies, but the comparatively higher rates in coarse-compared with fine-mesh litterbags in forest sites in this study ($kc/kf > 1.5$) indicate that shredders contributed to leaf breakdown. However, shredder contribution to leaf breakdown was dependent on leaf species suggesting that replacement of indigenous riparian vegetation with poorer quality *Eucalyptus* species has the potential to reduce nutrient cycling. If unchecked, riparian and watershed deforestation will shift the functioning of these streams with foodwebs becoming more reliant on autochthonous sources and microbial processing of leaf litter, which cannot support diverse consumers and complex food webs.

Chapter

5

Linkage between DOM composition and whole-stream metabolism in headwater streams influenced by different land use

Publication to be based on this chapter:
Masese FO, Gettel GM, Salcedo-Borda JS, Irvine K, McClain ME. Linkage between DOM composition and whole-stream metabolism in headwater streams influenced by different land use. *In Submission*

Abstract

This study assessed the influence of land use change on the composition of dissolved organic matter (DOM) and its links with ecosystem metabolism in upland Kenyan streams. A total of 50 sites in 34 streams spread out across a land use gradient were sampled for spatial variation in DOM composition. Whole-stream rates of GPP and ER were determined using the upstream-downstream (two stations) diurnal dissolved oxygen change technique at 10 streams in forest ($n = 5$) and agriculture ($n = 5$) land uses during dry and wet conditions. Water samples were analyzed for fluorescence and absorbance spectra of DOM, concentrations of dissolved organic carbon (DOC), nutrients and major ions. Agriculture increased conservative ions and nutrients but the concentration of DOC did not differ among land uses. Absorbance and fluorescence spectrophotometry of DOM indicated notable shifts in DOM composition along the land use gradient. Forest streams were associated with higher molecular weight and terrestrially derived DOM whereas agriculture streams were associated with photodegradation, autochthonously produced and low molecular weight DOM due to the open canopy. Streams draining mixed land use displayed intermediate characteristics that were largely driven by near-stream reach-scale influences. However, aromaticity ($SUVA_{254}$) was high at all sites irrespective of land use, indicating that soils in agricultural areas were sources of humic and high molecular weight DOM that is likely mobilized during tillage. Longitudinal stream-size-dependent shifts in DOM composition were noted but trends mimicked and bore an imprint of land use. Gross primary production (GPP) and ecosystem respiration (ER) where generally higher in agriculture streams favoured by the low canopy and higher nutrient concentrations compared with forest streams. Both GPP and ER responded to seasonality with higher rates during the dry season at both forest and agriculture streams. GPP also increased with stream size consistent with increased primary production and changes in DOM composition with opening of the canopy. There were links between GPP and ER rates and DOM composition whereby optical properties of DOM largely agreed with direct measures of ecosystem metabolism as demonstrated by relationships between DOM indices (fluorescence and freshness indices, $\beta{:}\alpha$, and FI) and GPP and ER rates. This is one of the first studies to link land-use change, organic matter dynamics and ecosystem metabolism in African tropical streams, extending the geography in which these global-change processes have been documented. The potential consequences of these changes in East Africa are significant because of the close linkages between ecosystem services and human wellbeing and livelihoods.

1. Introduction

Headwater streams influence the ecology and geomorphology of higher order streams as sources of water, sediments, organic matter, woody debris and nutrients (Wipfli et al., 2007; Lamberti et al., 2010). The contribution of headwater streams to the biogeochemical budgets of watersheds is also important, especially in terms of transfer and transformation of carbon, nitrogen and associated elements (Williamson et al., 2008). It is now recognized that streams are not just conduits of material delivery but also play a major role in the global carbon cycle (Cole et al., 2007; Battin et al., 2008; Aufdenkampe et al., 2011). One way to determine whether headwater streams work as sinks or sources of carbon is through the analysis of whole-stream ecosystem metabolism. Ecosystem metabolism integrates processes that control nutrient cycling and organic matter dynamics in streams and its determination is a good indicator of both ecosystem functioning and health (Young et al., 2008; Tank et al., 2010; Staehr et al., 2012).

Dissolved organic matter (DOM) is the most ubiquitous form of carbon in aquatic ecosystems and plays an important role in nutrient cycling (Bormann and Likens, 1967; Wetzel, 1992). In headwater streams, DOM is primarily derived from leaching of leaf litter, but this allochthonous resource is considered to be more refractory to bacterial growth (Meyer et al., 1998). On the other hand, DOM derived from primary production and microbial activities in streams is generally more labile and bioavailable (Farjalla et al., 2009). The application of spectrophotometric techniques to measure optical properties of DOM has enabled its characterization for different purposes, including differentiating among allochthonous, autochthonous, waste-derived and agricultural sources in aquatic ecosystems (Jaffé et al., 2008; Fellman et al., 2010).

Much of our understanding of the effects of land use change on organic matter dynamics in streams and rivers is based on studies of temperate ecosystems (see Staehr et al., 2012 and references therein). This is disproportionate considering that tropical streams and rivers transport >60% of the global riverine carbon (Ludwig et al., 1996; Schlünz and Schneider, 2000), and display higher rates of CO_2 flux than their temperate counterparts (Aufdenkampe et al., 2011). The limited number of studies on ecosystem metabolism and carbon cycling in tropical streams has hampered generalizations about determinants of ecosystem functioning and predicting impacts of regional and global human disturbances, notably land use change and climate change.

Catchment land use change and loss of riparian corridors to deforestation have had a disproportionate influence on the functioning of headwater streams mainly through changes in biological communities, organic matter dynamics, nutrient input and metabolism rates (Bilby and Bisson, 1992; Bernot et al., 2010). Because DOM quality is as important as DOM quantity in the understanding of carbon dynamics in streams (Battin et al., 2008; Tank et al., 2010), many studies have sought to link catchment land use and DOM quality (composition) in streams. The influence of agricultual land use on DOM composition has been uneiquivocal, although the responses in specific properties of DOM have been variable. For instance, increased contributions of microbially derived and structurally less complex DOM to agriculture streams have been reported (Wilson and Xenopoulos 2009; Williams et al., 2010). In contrast, Graeber et al. (2012) found the percentage of land under agricultural use to increase the amount of structurally complex and aromatic DOM in streams. These apparently divergent findings suggests that the proportion of agricultural land use within a catchment alone is unlikely to explain patterns in DOM composition in recipient streams. Land use history, soil type, tillage practice or technique, catchment topology, climate and alterations to hydrological residence time and flow paths, in addition to the uptake rates and provenance to metabolism of the different pools of DOM, have all been identified to contribute somewhat to DOM concentration and composition in streams (Ogle et al., 2005; Ewing et al., 2006; Cawley et al., 2014). However, despite the potential linkage between DOM composition and ecosystem metabolism in aquatic ecosystems (Cammack et al., 2004; Barrón et al., 2014), few studies have investiaged this linkage in streams (Halbedel et al., 2013; Kaushal et al., 2014). To improve our understanding and models of global carbon cycles, more data are needed on headwater streams across different biomes and climates (Battin et al., 2008). More studies that examine the spatio-temporal variability in DOM amount and composition and its relationship with ecosystem metabolism are also needed to improve understanding of the functioning of riverine ecosystems in the current times of change.

Studies on fluvial metabolism and its controls in African tropical streams are very limited. As is the case with temperate low order streams, dense canopy cover in African tropical streams in humid montane forests limits light availability and maintains low water temperature (Chapman and Chapman, 2003), conditions that can lead to net heterotrophy. However, land use change as a result of deforestation is widespread (Foley et al., 2005; Chapman and Chapman, 2003; Fidelis, 2014), resulting in increased water temperature, inputs of nutrients and sediments and reduced leaf litter input (Magana, 2001; Kasangaki et al., 2008; Masese et al., 2014a,b). These changes have implications on organic matter dynamics, light regimes, primary production and the diversity and composition of biological communities. Even with the limited data on direct measures of ecosystem metabolism in African tropical streams, indirect measures based on macroinvertebrate functional feeding groups in streams across a land use gradient from forestry to agriculture have reported accompanying shifts in ecosystem functioning from heterotrophy to increasing autotrophy (Masese et al., 2014a).

In this study, I determined possible effects of agricultural land use on DOM composition and ecosystem metabolism in headwater streams. The spatial variation in DOM concentration and composition was compared at 50 sites located among 34 streams draining different proportions of forest and agricultural land in the headwaters of the Mara River basin, Kenya. Following the strong temporal flow variations in tropical streams, I also examined seasonal variation in land use-related patterns of DOM amount and composition at 10 streams (5 forest and 5 agriculture) that were sampled during the dry and wet seasons. The same sites were also used for direct measurements of ecosystem (whole-stream) metabolism (gross primary production and ecosystem respiration). I also investigated possible linkages between DOM composition and ecosystem metabolism in the 10 streams. I hypothesized that (1) the land use change from forest to agriculture has affected DOM composition in agriculture streams, (2) agriculture streams have higher rates of ecosystem metabolism as a result of increased nutrient concentrations and reduced canopy cover and finally, (3) because DOM composition can influence its bioavailability for metabolism and, reciprocally, ecosystem metabolism can affect DOM composition, I hypothesized that such dependencies should lead to relationships between measures of ecosystem metabolism and DOM composition in streams draining different land uses. To better understand sources of DOM and its cycling in tropical streams, I also investigated the longitudinal variation in the composition of DOM as streams increase in size (Vannote et al., 1980).

2. Methods

2.1 Study Area

The study was conducted in mid-elevation (1900 m – 2300 m a.s.l) streams draining the western slopes of the volcanic Mau Escarpment, which forms part of the Kenyan Rift Valley. The streams form the headwaters of the Mara River that flows to Lake Victoria. The river drains the extensive tropical moist montane broadleaf Mau Forest Complex (MFC) that is a major water tower in Kenya (Figure 5.1). The MFC is the largest single block of montane forest in East Africa, but its present coverage is much fragmented and reduced because of excisions for human settlement, coniferous forest plantations and large-scale and small-scale cultivation of tea (Lovett and Wasser, 1993). However, some intact forest blocks remain and are protected as part of forest reserves and national parks. People living in the adjoining areas are involved in semi-intensive smallholder agriculture, characterized by cash crops (mainly tea), food crops (mainly maize, beans and potatoes) and animal husbandry. This has also resulted in the loss of indigenous riparian vegetation along agriculture streams and rivers where exotic *Eucalyptus* species dominate riparian vegetation. Moreover, with human population growth averaging 3% p.a., land use practices in the region are bound to intensify with remnant forest along river corridors likely to be cleared for farming and other uses.

Climate of the area is relatively cool and seasonal, characterized by distinct rainfall seasons and low ambient temperatures that fall below $10^{\circ}C$ during the cold months of January-February. Annual precipitation ranges from 1000 - 2000 mm. Dry conditions occur during January-March and two wet conditions during March-May and October-December, which are periods for the long and short rains, respectively. Temporal variability in wet and dry conditions is high due to other controlling factors, especially variable sea-surface temperatures in the Indian Ocean, which strongly influence the short rains (Black, 2005). Spatially, rainfall in the river basin varies as a function of elevation, with higher rains on the Mau Escarpment.

The region contains moderate levels of species richness and relatively low rates of endemism in comparison to other tropical eco-regions around the equatorial belt of Africa (Obati, 2007). MFC is part of Kenyan montane forests that are found in the central and western highlands. The volcanic mountains contain evergreen seasonal forests and evergreen forests. Vegetation patterns of the MFC

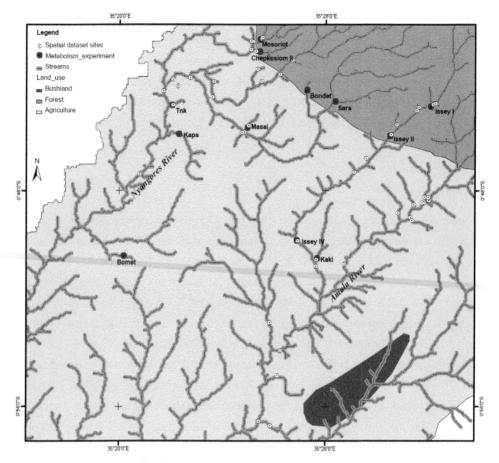

Figure 5.1. Location of metabolism sites and spatial dataset sites in the upper Mara River basin, Kenya.

are diverse, but there is a broad altitudinal zonation. Pockets of less-disturbed montane forest hold *Olea capensis*, *Prunus africana*, *Albizia gummifera* and *Podocarpus latifolius*, *Xymalos monospora*, *Syzigium guienense*, *Celtis africana*, *Aningeria adolfi-friederieri* and *Zanthoxylum gilletii* indicating fewer disturbances (Obati, 2007). *Croton macrostachyus* and *Syzygium cordatum* are common along riparian corridors of streams in the region and are among the ten most dominant species in the MFC and the surrounding areas (Blackett, 1994; Mathooko and Kariuki, 2000). Above 2300 m a.s.l the montane forest gives way to thickets of Bamboo *Arundinaria alpina* mixed with forest and grassland, and finally to montane sclerophyllous forest near the escarpment crest. Substantial parts of the high *Juniperus-Podocarpus-Olea* forest have been encroached and cleared, although sections remain in good condition. Cleared areas have been converted to plantation forest with pine and cypress species (Obati, 2007).

During the dry season shedding of leaves is higher in the region (Magana, 2001), and some tree species are known to shed all their leaves (pers. obser.). This phenological characteristic makes the

96

forest canopy to slightly open; it is much easier to walk through the forest and along streams for the same reason, as opposed to during the wet season when the vegetation is denser and undergrowth more impenetrable. Consequently shading in streams is minimized and insolation is higher and for longer periods due to limited cloud cover.

2.2 Sampling design

To capture spatial variability in dissolved organic matter characteristics, nutrient concentrations and water physicochemistry, 50 sites were selected along 34 streams on December 2011-January 2012. Sites were located in streams draining a gradient of catchment land use from 100% forestry to 100% agriculture and classified into three broad land use categories depending on the proportions of forest and agriculture land uses. Based on the Digital Elevation Model of Kenya (90 m by 90 m), obtained from the Shuttle Radar Topography Mission, catchments were delineated and the area of each land use category upstream of each sampling site calculated. Forest sites (FOR) and agriculture sites (AGR) drained catchments with the proportion of catchment land use under forestry and agriculture >70%, respectively. Mixed (MIX) sites did not meet the catchment land use criteria for FOR and AGR sites. Data collected from the 50 sites constitute the synoptic data set used to explore effects of catchment land use on organic matter characteristics, water quality, nutrient concentrations and DOM composition in AGR, MIX and FOR streams. In addition, ecosystem metabolism measurements and its links to DOM composition were conducted at 10 streams (5 AGR and 5 FOR) during the wet (November - December 2011) and dry (January - March 2012) seasons. These data constitute the experimental data set.

2.3 Physical and chemical variables

At the synoptic and experimental sites, pH, dissolved oxygen (DO), DO saturation, temperature and electrical conductivity were measured *in situ* using a YSI multi-probe water quality meter (556 MPS, Yellow Springs Instruments, Ohio, USA). Turbidity was measured using a portable Hach turbidity meter (Hach Company, 2100P ISO Turbidimeter, USA). Water samples were collected from the thalweg using acid washed HDP bottles for analysis of nutrients, major anions and cations, dissolved organic carbon (DOC) and particulate organic matter (POM). For total suspended solids (TSS) and POM, water samples were filtered immediately through pre-weighed and pre-combusted (450°C for 4 h) GF/F filters (Whatman International Ltd., Maidstone, England). All samples were stored and transported in a cooler to the laboratory and frozen within 10 hours of sampling.

Water samples for DOM characterization were collected in 30 ml amber glass bottles. The bottles were cleaned with 0.1M HCl, soaked overnight in distilled water and then combusted (450°C, 4 h) before transport to the field. Water samples were filtered on site using pre-combusted (500 °C, 4 h) GF/F filters. For the synoptic data set replicate samples were taken per site. For the experimental data set, samples were taken from upstream and downstream of each study reach. Samples were wrapped in aluminium foil and transported in a cooler before being frozen in the laboratory until analysis. Freezing of samples was necessary because of the long period of storage of samples before analysis in Delft, the Netherlands. However, freezing has been shown to reduce specific ultraviolet absorbance (SUVA), total dissolved phosphorus and DOC concentration (Spencer et al., 2007; Fellman et al., 2008). Short-term cold storage in a refrigerator has similar effects as single freezing and thawing (Hudson et al., 2009), hence stronger long-term effects of cold storage due to continuous oxidation and microbial activity that would occur were potentially avoided by freezing and thawing the samples only once and processing them identically. It is therefore assume that this impacted all samples similarly and had minimal influence on the results.

2.4 Ecosystem metabolism

Whole-stream rates of GPP and ER were determined using the upstream-downstream diurnal dissolved oxygen (DO) change technique (Marzolf et al., 1994; Young and Huryn, 1998) in a 100 m study reach in each stream. This method allows metabolism estimates in a parcel of water flowing between two points in a reach by attributing changes in O_2 to photosynthesis, respiration and re-aeration. DO and temperature were measured using Hydrolab sondes (MS5 equipped with luminescent dissolved oxygen sensors, Hach Hydromet) that were set to record data every 5-20 minutes at upstream and downstream stations over a 24-h period. Oxygen flux was calculated based on the average oxygen saturation deficit or excess within the study reach. The re-aeration rate (k) was measured from changes in dissolved propane concentration during steady-state injection of propane and a conservative tracer (Cl-) used to account for dilution of propane caused by groundwater inflow (Genereux and Hemond, 1992). The re-aeration rate of propane was converted to oxygen using a factor of 1.39 (Rathbun et al., 1978). For comparison, k was also calculated using the physical characteristics of the stream channel using the energy dissipation model (EDM, Tsivoglou and Neal, 1976) as follows: k_{20} = K' x S x V where k_{20} is the oxygen re-aeration rate at 20 °C (day^{-1}), K' is an empirical constant equivalent to 28.3 x 10^3 s m^{-1} day^{-1} for streams with discharge values < 280 L s^{-1}, S is the channel slope (m m^{-1}) and V is velocity (m s^{-1}). k obtained from the two methods were corrected for water temperature according to Elmore and West (1961). Although there are a number of physically based methods for estimating re-aeration rate (Genereux and Hemond, 1992), I chose EDM because it has been recommended for use in open-system methods for determining metabolism in streams (APHA, 1998).

Gaseous O_2 exchange with the atmosphere (ΔDO) was calculated based on the oxygen saturation deficit or excess within the study reach and corrected for re-aeration (k) following Marzolf et al. (1994): M = ((C_t - C_{to})/Δt – $K_{O2}D$)*Z, where C_{to} is the O_2 concentration at the upstream site (g O_2 m^{-3}), C_t is the O_2 concentration at the downstream site, Δt is the travel time, K_{O2} is the temperature-corrected re-aeration coefficient of O_2, D is the saturation deficit or excess and Z is mean stream depth. Ecosystem respiration (ER) was calculated by summing the re-aeration-corrected ΔDO measured during the night while daytime ER was determined by extrapolating between the net oxygen change rate during the 1-h predawn and postdusk periods. Gross primary production (GPP) was calculated by summing the respiration rate during the photoperiod and ΔDO. GPP and ER were used to calculate net ecosystem production, which reflects the balance between autotrophic and heterotrophic processes in the ecosystem, as the difference between GPP and ER (i.e., NEP = GPP – ER).

A number of physical, chemical and biological characteristics were determined at each study reach to establish relationships with stream metabolism. Canopy cover above each stream was estimated visually and expressed as a percentage. Stream width, depth and water velocity were measured at 11 transects located along each reach, and discharge was estimated using the velocity-area method. Known volumes of water were filtered through a 0.7 μm pore-sized GF/F filters for water column chlorophyll *a* determinations. In addition, samples from each major benthic substrate type were collected for benthic chlorophyll *a* determination. For hard surfaces (e.g. cobbles), a recorded area of substrate was scrubbed for biofilm and the slurry filtered through a 0.7 μm GF/F filtrers. For soft sediments (gravel, sand and silt), a fixed area of the top 20mm of substrate was removed using a cut-off 60ml syringe. All chlorophyll *a* samples were wrapped in aluminium foil to prevent exposure to light, transported on ice and stored frozen in the laboratory until analysis. Triplicate samples of coarse particulate organic matter (CPOM) were collected from pools, riffles and runs by kicking a standard $1m^2$ of stream bottom using a kick net (mesh size 1 mm). The standing crop of detrital fine benthic organic matter (FBOM) was determined according Mulholland et al. (2000). An open-ended bucket was placed into the stream and sediments vigorously agitated to a depth of about 10 cm. The slurry was subsampled using 500 ml HDEP bottles. Because of logistical constraints, comprehensive data on particulate organic matter quantity and instream characteristics were collected only during the dry season.

2.5 Laboratory analyses

Alkalinity, total dissolved phosphorus (TDP), total phosphorus (TP), soluble reactive phosphorus (SRP) and total nitrogen (TN) were determined using standard colorimetric methods (APHA, 1998). GF/F filters holding suspended matter were dried (95°C) to constant weight and TSS was determined by re-weighing on an analytical balance and subtracting the filter weight. The filters were then ashed at 500°C for 4 h and re-weighed for determination of POM as the difference between TSS and ash-free-dry weight. Dissolved organic carbon (DOC) and total dissolved nitrogen (TDN) concentrations were determined using a Shimadzu TOC-V-CPN with a coupled total nitrogen analyzer (TNM-1) and used to calculate C:N ratio. Dissolved organic nitrogen (DON) was calculated by subtracting the inorganic nitrogen (NO_3-N and NH_4^+) from TDN. Chlorophyl a pigments were extracted by 90% ethanol and concentrations were determined spectrophotometrically (APHA, 1998). Major anions NO_3^-, ortho-phosphate (PO_4^{3-}), Cl^- and SO_4^{2-} were determined using a Dionex ICS-1000 ion chromatographer equipped with an AS-DV auto sampler, and the major cations Na^+, K^+, Ca^{2+}, Mg^{2+}, dissolved silicates (DSi) and NH_4^+ using an ICP-MS. CPOM samples were sorted to remove invertebrates and inorganic materials and dried to a constant mass at 68°C for 48 h. The mass of different CPOM fractions- leaves, sticks, seeds and flowers - were weighed separately using a Sartorius balance (precision 0.1 mg). The FBOM samples were dried (68 °C), then weighed, combusted (500 °C) and reweighed to determine AFDM. CPOM and FBOM biomass were expressed per unit area sampled.

2.5.1 Optical properties of DOM

Absorption spectra (200 to 600 nm) of DOM were measured on a UV-2501PC UV/VIS spectrophotometer (Shimadzu, Duisburg, Germany) using a 1 cm quartz cuvette. Prior to analysis, samples were brought to room temperature. MilliQ-Water was used as a blank. Absorption coefficients were determined following $a\lambda = 2.303$ $A(\lambda)$ / l where $A(\lambda)$ is the absorption coefficient at wavelength λ (in nm) and l the cuvette path length (m). A number of optical properties of DOM were calculated from the scans. The absorption coefficient ratio a_{254}/a_{410} was calculated as an indicator of molecular weight and aromaticity (Baker et al., 2008). The specific UV absorbance at 254 nm ($SUVA_{254}$) which is the ratio between the UV absorbance at 254 nm (m^{-1}) and DOC concentration (mg L^{-1}) was calculated as an indicator of aromaticity and DOM concentration (Weishaar et al., 2003). A commonly used ratio of absorption coefficients $E2:E3$ ($a_{250}:a_{365}$) was calculated and used to provide further information about DOM aromaticity and molecular weight (Peuravuroi and Pihlaja, 1997; Helms et al., 2008). The spectra slope ratio (S_R) was computed as the ratio of the short wavelength slope ($S_{275-295}$) and the long wavelength slope ($S_{350-400}$). Both S_R and $S_{275-295}$ are inversely correlated with average molecular weight of DOM and are associated with photodegradation (Helms et al., 2008).

Fluorescence of DOM was measured with a FluoroMax-3 spectro-fluorometer (Jobin Yvon [now HORIBA Scientific], Longjumeau, France). Excitation-emission matrices (EEMs, Figure 5.2) were obtained by a 3D-scan of fluorescence over an excitation range of 220 to 450 nm (at 10 nm increments) and at an emission range of 350 to 600 nm (at 2 nm increments) using methods outlined in Cory and McKnight (2005) and Cory et al. (2010). To correct for instrument bias related to wavelength-dependent efficiencies, instrument specific files supplied by the manufacturer were applied. Normalized blank EEMs were subtracted from each sample EEM to eliminate effects of Raman and Rayleigh scattering. EEMs were also corrected for inner-filter effects and normalized to Raman units (in nm^{-1}, Raman peak area of the blank at 350 nm excitation). The processing of EEMs

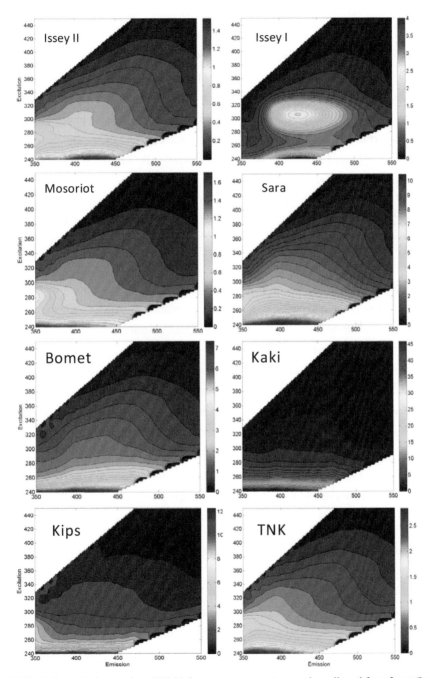

Figure 5.2. Excitation-emission matrices (EEMs) from wet season water samples collected from forest (Issey II, Issey I, Mosoriot and Sara) and agriculture (Bomet, Kaki, Kips and TNK) streams in the upper Mara River basin, Kenya.

101

was done in MATLAB 8.2 (Mathworks, MATLAB 2013) and yielded three fluorescence indices: fluorescence index (FI) (McKnight et al., 2001); freshness index ($\beta:\alpha$) (Wilson and Xenopoulos, 2009) and biological autochthonous index (BIX) (Huguet et al., 2009). The FI provides information on DOM origin, distinguishing terrestrially derived DOM (FI~1.3) from microbially derived DOM (FI~1.8) (McKnight et al., 2001). $\beta:\alpha$ indicates the proportion of recently produced DOM relative to more decomposed DOM (Parlanti et al., 2000; Wilson and Xenopoulos, 2009). $\beta:\alpha$ values > 1 indicate that DOM is primarily of autochthonous origin and values < 0.6 indicate primarily allochthonous origin (Huguet et al. 2009). BIX estimates autochthonous biological activity with values higher than 0.8 indicating freshly released and autochthonously produced DOM whereas lower values indicate less autochthonous DOM (Huguet et al., 2009).

2.6 Statistical analysis

For the synoptic data set, one-way analysis-of variance (ANOVA) was used to test for differences in water quality variables, optical properties of DOM and its concentration among land uses followed by *post hoc* Tukey's Honestly Significant Difference (HSD) multiple comparisons of means. Principal Component Analysis (PCA) was used to summarize variation in physico-chemical variables and optical properties of DOM among land uses. Effects of land use and river distance (RDS) were tested on the first four principal components arising from the PCAs using analysis-of-covariance (ANCOVA) with land use, percent of land use under agriculture (arcsine of % AGR) and RDS (calculated as the as √drainage area [DA]) as covariates. The length of stream paths leading to a point in the drainage can be expressed as a power function of the DA which has been estimated to be 0.5 (Gregory and Walling, 1973). Rasmussen et al. (2009) notes that √DA represent the average distance covered by tributaries before they join at a point in the mainstem. Thus, √DA was used both as a measure of a size of a stream and its longitudinal distance. Consequently, to more closely examine the effect of RDS on DOM composition, relationships between optical properties of DOM and RDS were explored by pair-wise simple linear regression (SLR).

For experimental data set, PCA was used to condense and summarize data that were grouped into four multivariate data sets; (i) water physico-chemistry including conservative ions and nutrients (but excluding all carbon-related information), (ii) all absorbance and fluorescence indices (optical properties) describing the quality of DOM, (iii) all data related to organic matter quantity (both dissolved and particulate), including TSS and canopy cover, and (iv) measures of stream size (catchment area, discharge, depth, wetted and flowing width). The first four PCA-axes (principal components) of the multivariate data sets were subsequently used as factors controlling rates of GPP and ER in simple and stepwise multiple regressions. Other factors considered were Ln TDN, Ln NO_3-N, Ln NH_4^+, Ln DOC, Ln FBOM, Ln CPOM, arcsine % canopy cover, arcsine % AGR land use, Ln C:N ratio and arcsine % POM in TSS. Correlation analysis was used to examine the association

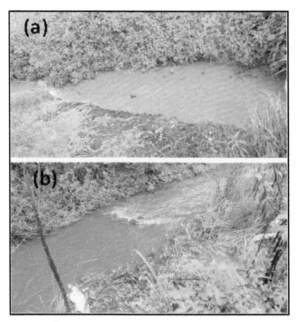

Plate 5.1: Effects of runoff on suspended sediments in (a) agriculture and (b) forest streams in the upper Mara River basin. The photos were taken minutes after a rain event.

between rates of GPP and ER. Simple linear regressions were used to examine relationships between metabolism measurements and DOM composition by season. Both the individual optical indices (absorbance and fluorescence based) and the first four PCA-axes of DOM composition were considered. Mann-Whitney U tests (Mann and Whitney, 1947) were used to test for differences in GPP and ER rates between season and land uses. Statistical analyses and graphs were done in Statistica (Statistica, Version 7, StatSoft, Tulsa, Oklahoma), SigmaPlot Version 12.0 (Systat Software, San Jose, CA) and Microsoft Excel 2007. Data were appropriately transformed using natural-log or arcsine transformations before analysis to meet assumptions for parametric tests.

3. Results

4. Land use effects on water quality and DOM composition

Synoptic data set: Streams in agricultural (AGR) and forest (FOR) catchments showed differences in physico-chemistry and DOM composition (Plate 5.1, Table 5.1). Mixed streams (MIX) displayed no particular pattern in some of the variables and behaved as either AGR or FOR streams or both. Following PCA ordinations to establish land use effects on water quality and DOM composition (Figure 5.3abcd), Factor 1 (PCA-axis 1) accounted for most variation (52.8%) in the water quality PCA ordinations. Higher concentrations of nutrients (TDN and NO_3-N) and conservative ions (Cl$^-$, SO_4^{-2}) were associated with agriculture streams (Figure 3ab). Land use driven variation of DOM

Table 5.1. Land use affects stream water nutrients, conservative ion concentrations and optical properties of DOM and organic matter. Mean (±SD) physico-chemical water quality variables and optical properties of dissolved organic matter and its concentration, carbon to nitrogen (C:N) ratio, organic matter standing stocks and chlorophyll *a* concentrations for three catchment land use categories mixed (MIX), forest (FOR) and agriculture (AGR) in the upper Mara River basin, Kenya. Statistics of one-way analysis of variance (ANOVA) results among three land use categories are presented together with *post hoc* Tukey's HSD comparisons among means. Similar superscripts among means indicate lack of significant differences between land uses at $p < 0.05$. Model degrees of freedom (df) = 2 and residual df = 31. *asterisk indicate significant differences at $p < 0.05$, one-way ANOVA. [#]

	Land use means ± SD			Statistics	
	FOR	MIX	AGR	F	*p*
RDS (sqrt DA in Ha)	38.1 ± 21.8^a	206.6 ± 69.0^b	19.7 ± 18.0^a	67.5	0.0000*
DOC (mg/L)	1.7 ± 0.4^a	1.1 ± 0.4^a	1.4 ± 0.7^a	3.0	0.0642
TDN (mg/L)	1.2 ± 0.5^a	1.7 ± 1.2^a	6.6 ± 2.6^b	44.3	0.0000*
TDON (mg/L)	0.2 ± 0.1^a	0.2 ± 0.2^a	0.5 ± 0.2^b	8.8	0.0009*
Dissolved C:N[#]	10.2 ± 4.0^a	8.5 ± 6.9^a	4.0 ± 2.6^b	9.3	0.0007*
TDP (mg/L)	0.05 ± 0.06^a	0.03 ± 0.01^a	0.03 ± 0.01^a	2.2	0.1238
Chloride (mg/L)	3.2 ± 0.7^a	2.7 ± 0.7^a	6.3 ± 3.4^b	11.5	0.0002*
NO_3-N (mg/L)	1.0 ± 0.4^a	1.5 ± 1.1^a	6.1 ± 2.6^b	46.1	0.0000*
SO_4^{-2} (mg/L)	2.6 ± 0.2^a	1.8 ± 0.6^b	4.6 ± 3.3^a	7.0	0.0031*
NH_4-N (mg/L)	0.02 ± 0.01^a	0.02 ± 0.01^a	0.04 ± 0.03^b	6.2	0.0053*
FBOM (g/m²)	515 ± 570^a	629 ± 709^a	1203 ± 1090^a	2.3	0.0813
CPOM (g/m²)	145.6 ± 13.6^a	169.1 ± 19.2^b	233.0 ± 14.4^b	8.4	0.0410*
Water column chlorophyll *a* (µg/L)	6.1 ± 6.5^a	9.8 ± 7.1^a	37.5 ± 21.1^a	3.9	0.0303*
Benthic chlorophyll a (µg/M²)	2.7 ± 1.5^a	14.3 ± 4.8^a	17.5 ± 11.1^a	7.7	0.0031*
TSS (mg/L)	36.6 ± 20.5^a	173.5 ± 14.6^b	254.9 ± 74.1^b	20.21	0.0001*
% carbon in TSS	13.2 ± 4.2^a	8.7 ± 6.4^{ab}	4.1 ± 3.2^a	7.6	0.0023
$SUVA_{254}$	24.3 ± 9.9^a	28.0 ± 4.9^a	42.3 ± 18.5^b	8.5	0.0012*
a_{254}/a_{410}	6.2 ± 4.9^a	5.3 ± 0.5^{ab}	4.9 ± 0.8^b	7.2	0.0027*
FI	1.3 ± 0.14^a	1.4 ± 0.06^b	1.4 ± 0.07^b	6.3	0.0051*
β:α	0.4 ± 0.3^a	0.9 ± 0.1^b	0.7 ± 0.1^b	8.9	0.0009*
BIX	0.7 ± 0.1^a	0.8 ± 0.3^{ab}	0.9 ± 0.1^b	7.1	0.0029*
Total absorbance (200-600 nm)	11704.5 ± 3081.5^a	10782.5 ± 4561.8^a	20359.4 ± 9927.1^b	9.9	0.0005*
S_R	1.2 ± 0.1^a	1.2 ± 0.1^a	1.4 ± 0.2^b	8.9	0.0009*
$S_{275-295}$	0.013 ± 0.001^a	0.013 ± 0.001^a	0.013 ± 0.001^a	1.8	0.1828
E2:E3	3.8 ± 0.2^a	3.7 ± 0.4^a	4.0 ± 0.4^a	3.0	0.0640

[#]Note: RDS = river distance expressed as the square root (sqrt) of drainage area (DA) in hectares (Ha). Dissolved C:N was calculated as (DOC/12000)/(TDN/14000). FBOM, water column chlorophyll a and benthic chlorophyll a data were collected mainly from experimental sites used for metabolism measurements. CPOM data was obtained during the dry season – for details see Chapter 4.

composition was visible along the first and second PCA-axes (Factor 1 and 2) with most FOR and AGR sites separated along PCA-axis 1 (44.9%) and MIX sites lying intermediate (Figure 5.3cd). However, the separation was not complete along PCA-axis 2, implying that despite the catchment land use influences reach-scale influences also affected some DOM properties. AGR sites were associated with higher proportions of recently produced (β:α), autochthonously produced (BIX) DOM, aromaticity (SUVA$_{254}$) and the spectra slope ratio (S$_R$) (Table 5.1).

Using PCA-scores derived from water quality (WQ) variables as dependent variables in ANCOVA, there were significant effects ($p<0.05$) of land use and the proportion of agricultural land use (arcsine % GAR) and river distance on WQ-PC 1 only (Table 5.2). For DOM composition, a significant land use effect occurred along DOM-PC 2 and DOM-PC 3 whereas arcsine % AGR had a significant effect on DOM-PC 1 and DOM PC 4. River distance had a significant effect ($p<0.05$) along DOM-PC 4 only. WQ-PC 1 displayed a significant inverse relationship with DOM-PC 1 (SLR, R^2_{adj} = 0.23, p < 0.01) and DOM-PC 3 (R^2_{adj} = 0.34, p < 0.001), but since DOM-PC 1 and DOM-PC 3 are orthogonal axes of a PCA, they captured different aspects of variability in DOM composition. For instance, only DOM-PC 1 was related with arcsine % AGR (R^2_{adj} = 0.21, p < 0.01). This implies that DOM-PC3 responded more to catchment-scale driven influences such as nutrient loading and declines in water quality, while DOM-PC1 responded more to reach-scale influences such as loss of canopy cover and nutrient loading. The other important DOM-PC axes (DOM-PC 2 and DOM-PC 4) were not related to any WQ-PC-axes. However, DOM-PC 2 responded to land use indicating similar responses to DOM-PC 3. Similar to DOM-PC 1, DOM-PC 4 responded to arcsine % AGR suggesting that shifts along this axis could have been driven by increases in primary production and photodegradation as a result of reduced canopy cover.

Based on the SLR models used to test for longitudinal changes in DOM composition, aromaticity (SUVA$_{254}$), short wavelength slope (S$_R$), and total absorbance (200-600 nm) reduced with river distance, whereas the proportion of recently produced DOM (β:α) and biological-autochthonous index (BIX) increased with river distance (Figure 5.4) consistent with increased autochthonous DOM production as streams widen. Microbially derived DOM (FI), molecular weight (a_{254}/a_{410} and E2:E3) and short wavelength slope (S$_{275-295}$) did not show any relationships with river distance. However, β:α, S$_R$ and FI are highly correlated (Figure 5.3) and have higher values in AGR streams (Table 5.1).

Experimental data set: Similarly to the synoptic data set, AGR sites were associated with higher electrical conductivity, temperature and concentrations of dissolved nutrients (NH$_4$-N, NH$_3$ and NO$_3$-N) and major ions (Cl$^-$, Ca^{+2}, mg^{+2} and SO$_4^{-2}$) (Figure 5.5a,d). Land use influenced changes on organic matter quantity along the first PCA-axis (OM-PC1, Figure 5.5b). On the second PCA-axis (OM-PC2) the separation was also significant with four of the five AGR sites and three of the five FOR sites

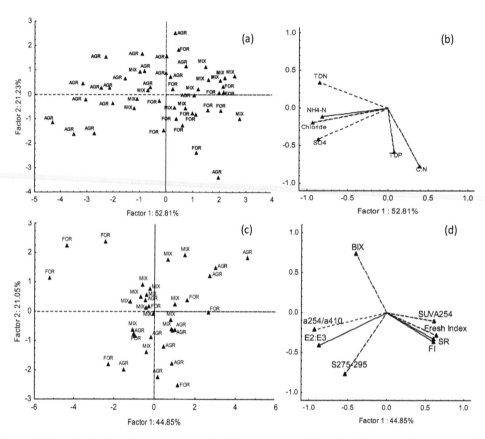

Figure 5.3. Results of principal component analysis (PCA) on water quality and DOM composition of the spatial dataset. (a) shows the scores for the land uses based on water quality and (b) shows the loadings of the water quality data (n=47). (c) shows the scores for the land uses based on the optical properties of DOM and (d) the loadings of the optical properties of DOM (n=35); Fresh Index= freshness index (β:α).

Table 5.2. Results of analysis of covariance (ANCOVA) showing effects of land use, % agricultural land use (% AGR) and river distance (RDS, sqrt of drainage area in hectares) on PCA combined response matrices based on water quality (WQ-PC axes) and optical properties of DOM (DOM-PC axes) on the first four principal components. * = p < 0.05.

Source of variation	df	MS	F	p
WQ- PC 1				
Land use	1	87.65	46.60	<0.001*
Arcsine of % AGR	1	22.95	12.20	0.001*
RDS	1	17.13	9.11	0.004*
Error	43	1.88		
WQ-PC 2				
Land use	1	0.02	0.02	0.900
Arcsine of % AGR	1	4.88	3.31	0.076
RDS	1	1.178	0.80	0.377
Error	43	1.48		
WQ- PC 3				
Land use	1	0.68	0.68	0.414
Arcsine of % AGR	1	2.99	2.98	0.092
RDS	1	0.37	0.378	0.546
Error	43	1.01		
WQ-PC 4				
Land use	1	0.341968	0.838	0.367
Arcsine of % AGR	1	0.355930	0.868	0.358
RDS	1	0.159968	0.39	0.537
Error	43	0.412007		
DOM- PC 1				
Land use	1	3.00	1.56	0.221
Arcsine of % AGR	1	22.45	11.66	0.002*
RDS	1	0.0002	0.0001	0.991
Error	31	1.93		
DOM-PC 2				
Land use	1	8.88	6.35	0.017*
Arcsine of % AGR	1	0.72	0.51	0.480
RDS	1	1.96	1.40	0.245
Error	31	1.40		
DOM-PC 3				
Land use	1	13.49	15.72	<0.001*
Arcsine of % AGR	1	2.34	2.72	0.109
RDS	1	0.004	0.004	0.948
Error	31	0.86		
DOM-PC 4				
Land use	1	0.52	1.67	0.206
Arcsine of % AGR	1	2.43	7.79	0.009*
RDS	1	1.43	4.58	0.040*
Error	31	0.31		

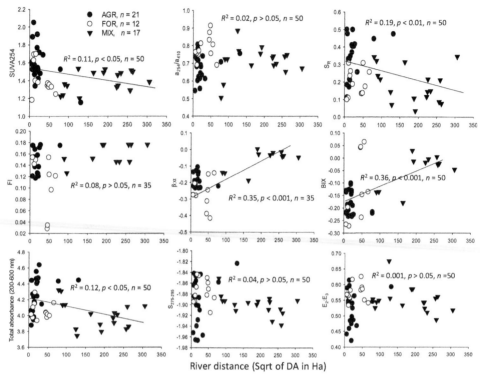

Figure 5.4. Longitudinal changes in log-transformed optical properties of dissolved organic matter as determined by regression models between river distance as an independent predictor and optical properties of DOM as dependent variables.

separated along this axis. AGR sites were associated with higher concentrations of DOC, benthic and water column chlorophyll a, TSS and lower C:N ratio (also Table 1). FOR sites were associated with higher biomass of CPOM, canopy cover, proportion of carbon (OM) in TSS (% POM) and C:N ratio. There were no clear distinctions between AGR and FOR streams in terms of stream size characteristics (Figure 6e).

Land use-associated variation of DOM composition was visible along the first PCA-axis (DOM-PC1) during the dry season (Figure 5.5c). Two FOR streams, which were also the widest (see Table 3 for details), were associated with autochthonously produced DOM (BIX), while the rest were associated with aromaticity (SUVA$_{254}$) and higher molecular weight of DOM (a$_{310}$/a$_{410}$, E2:E3). AGR sites were associated with photodegradation (S$_R$) and recently produced DOM (β:α). Trends were unclear during the wet season as PCA failed to identify land use-linked influences on DOM composition for many of the sites (Figure 5.5f). Both AGR and FOR streams had a mixture of DOM sources with both lower ($E2:E3$ and S$_R$) higher molecular weight DOM and aromaticity (a$_{254}$/a$_{410}$).

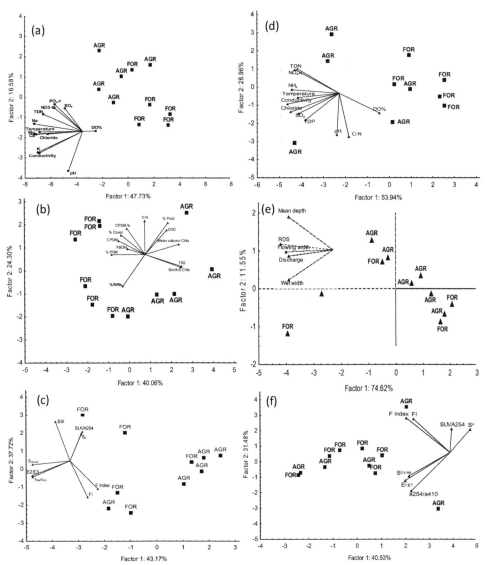

Figure 5.5. Separation of land use influences in PCA on the experimental data set based on (a, d) water physico-chemistry, including major ions and nutrients, (b) measures of organic matter quantity, (e) stream size variables, and (c, f) optical properties of dissolved organic matter, during the dry (a, b, c, e) and wet (d, f) seasons. Variation explained by each PC-axis, (Factor 1 and 2) is indicated by the regression coefficient (R^2) which is expressed as a percentage. F index = freshness index (also β:α), FI = fluorescence index.

3.3 Whole-stream metabolism

The location, drainage area, stream order and physical characteristics of experimental streams in which whole-stream metabolism experiments were conducted are presented in Table 5.3. Agriculture streams had lower canopy levels and higher water temperature than forest streams. Most streams were net heterotrophic (GPP/ER < 0.5 g O_2 m^{-2} day^{-1}) during both the dry and wet seasons (see details in discussion, Table 5.4. Dry season GPP and ER rates were higher than wet season rates for most streams. Lowest GPP and ER rates were measured in FOR streams and highest rates in AGR streams, although there were some overlaps. However, GPP rates in all FOR streams were lower (range 0.14-1.66 g O_2 m^{-2} day^{-1}) than the lowest GPP rate (3.59 g O_2 m^{-2} day^{-1}) in an AGR stream. The lowest GPP rates in FOR streams were 0.13 and 0.14 g O_2 m^{-2} day^{-1} during the wet and dry seasons, respectively (Table 5.4). Comparatively, the lowest GPP rates in AGR streams, 0.71 and 3.59 g O_2 m^{-2} day^{-1} during the wet and dry seasons respectively, were more than five times higher than the lowest GPP rates in FOR streams. Similarly, the highest GPP rates in AGR streams were more than four times the highest GPP rates in FOR streams during both the dry and wet seasons.

The ER rates were highly variable among streams within similar land uses. ER rates ranged from -1.04 to -4.83 and -4.52 to -8.09 g O_2 m^{-2} day^{-1} at FOR streams during the wet and dry seasons, respectively. In AGR streams ER rates ranged from -2.88 to -49.17 and -4.52 to -56.87 g O_2 m^{-2} day^{-1} during the wet and dry season, respectively. There was a significant, positive correlation between GPP and ER (r = 0.86, p < 0.0001, n = 20) and the correlation was stronger during the wet (r = 0.91, p < 0.001, n = 10), compared with the dry season (r = 0.74, p = 0.015, n = 10).

There were significant differences (Mann-Whitney U= 22.5, p = 0.038) in GPP rates between the dry (Mean±SD, 4.7±5.1 g O_2 m^{-2} day^{-1}) and wet (2.3±4.3 g O_2 m^{-2} day^{-1}) seasons. Similarly, ER rates also differed ((Mann-Whitney U= 20.0, p = 0.023) between the dry (-15.1±16.9 g O_2 m^{-2} day^{-1}) and wet (-8.4±14.5 g O_2 m^{-2} day^{-1}) seasons. There were significant difference in GPP rates between FOR and AGR streams during the dry (Mann-Whitney U= 15.0, p = 0.009) and wet (Mann-Whitney U= 15.0, p = 0.009) seasons, but ER rates did not differ. The same AGR site (Bomet) recorded the highest rates of GPP and ER rates during the wet (14.39 and 49.17, respectively) and dry (15.34 and 56.87) seasons.

There were no significant differences in NEP between the dry (Mean ± SD, -10.3±12.2) and wet (-6.2±10.2) seasons (Mann-Whitney U= 29.0, p = 0.112) and between forest and agriculture land uses during the dry (Mann-Whitney U= 9.0, p = 0.465) and wet (Mann-Whitney U= 9.0, p = 0.548) seasons. During the wet season NEP in forest streams ranged from -4.31 g O_2 m^{-2} day^{-1} to -0.91 g O_2 m^{-2} day^{-1} (P/R range: 0.09–0.17) . During the same period NEP in agriculture streams ranged from -34.78 g O_2 m^{-2} day^{-1} to -1.41 g O_2 m^{-2} day^{-1} (P/R range: 0.17–0.51). During the dry season NEP in forest streams ranged from -7.95 g O_2 m^{-2} day^{-1} to -2.86 g O_2 m^{-2} day^{-1} (P/R range: 0.02–0.37) while

Table 5.3. Location, drainage area, stream order and physical characteristics of experimental streams in agriculture and forest catchments in the upper Mara River basin, Kenya; [¥]identifies forest streams, [#]sampled during the wet and [β]sampled during the dry season only.*

Site	Season	Latitude (S)	Longitude (E)	DA (Ha)	Stream order	Gradient (m m^{-1})	k (measured)	k (calculated)
Issey I[¥]	Dry	0° 44' 48.8"	35° 32' 07.7"	2320	2nd	0.015	34.9	35.4
	Wet							39.6
Issey II[¥]	Dry	0° 49' 47.1"	35° 26' 53.1"	3198	3rd	0.019	29.6	32
	Wet							44.9
Bondet[¥]	Dry	0° 44' 27.7''	35° 27' 15.0''	207	1st	0.038	29.4	29.9
	Wet							39.6
Mosoriot[¥]	Dry	0° 42' 18.2''	35° 25' 28.8''	455	1st	0.025	4.1	13.1
	Wet							19.5
Sara[¥]	Dry	0° 44' 41.1''	35° 28' 24.9''	996	1st	0.021	28.3	24.8
	Wet							27.73
Masai[β]	Dry	0° 45' 36.6''	35° 25' 0.3''	202	1st	0.022	30.9	47.2
Issey IV[#]	Wet	0° 46' 50.7''	35° 26' 55.1''	5985	3rd			45.4
Kaps	Dry	0° 44' 46.9''	35° 22' 06.7''	236	1st	0.025	37	39.3
	Wet							43.6
Bomet	Dry	0° 47' 47.9''	35° 20' 05.5''	1398	2nd	0.022	2	3.7
	Wet							25.9
Kaki	Dry	0° 50' 34.6''	35° 27' 35.1''	698	1st	0.021	3.5	9.2
	Wet							14.8
TNK	Dry	0° 45' 53.7''	35° 22' 20.1''	1270	2nd	0.021	51.6	33
	Wet							39.5

Table 5.3 continued

Site	Season			Discharge (L/s)	Wetted width (m)	Depth (m)	% canopy cover	Temperature (°C)
Issey I[¥]	Dry	0° 44' 48.8"	35° 32' 07.7"	64.4	3.1	0.17	55	12.9
	Wet			187.6	3.4	0.28	70	14.9
Issey II[¥]	Dry	0° 49' 47.1"	35° 26' 53.1"	71.8	4.1	0.15	35	16
	Wet			119.9	4.8	0.16	40	15.3
Bondet[¥]	Dry	0° 44' 27.7''	35° 27' 15.0''	3.1	2.4	0.09	70	14.4
	Wet			58.6	2.5	0.22	85	14.3
Mosoriot[¥]	Dry	0° 42' 18.2''	35° 25' 28.8''	2.65	1.8	0.08	70	13.1
	Wet			22.2	2.3	0.12	90	14.2
Sara[¥]	Dry	0° 44' 41.1''	35° 28' 24.9''	18.2	1.7	0.15	70	14.1
	Wet			57.7	1.9	0.19	75	14.3
Masai[β]	Dry	0° 45' 36.6''	35° 25' 0.3''	6.7	2	0.08	40	16.6
Issey IV[#]	Wet	0° 46' 50.7''	35° 26' 55.1''	116.9	3.7	0.27	50	17.6
Kaps	Dry	0° 44' 46.9''	35° 22' 06.7''	21.7	1.8	0.18	35	19.1
	Wet			62.8	2.2	0.15	40	18.2
Bomet	Dry	0° 47' 47.9''	35° 20' 05.5''	2.7	1.6	0.15	35	17.6
	Wet			189.6	2.1	0.29	40	17.8
Kaki	Dry	0° 50' 34.6''	35° 27' 35.1''	3.5	1.2	0.11	25	18.8
	Wet			54.9	1.9	0.15	30	18.3
TNK	Dry	0° 45' 53.7''	35° 22' 20.1''	13.6	1.5	0.09	45	18.1
	Wet			116.5	1.7	0.28	50	18.6

*Note: k (measured) was determined from the propane evasion method only during the dry season, while k (calculated) was determined using the energy dissipation model during both the dry and wet seasons. Discharge, wetted width, depth and % canopy cover values are means calculated from transects throughout the study reach during both the dry and wet seasons. Temperature values are means of logged data (Hydrolab, MS5 sondes) from upstream and downstream stations during the 24-h whole-stream metabolism period.

Table 5.4. Dry and wet season gross primary production (GPP), ecosystem respiration (ER) rates and net ecosystem production (NEP) in streams in agriculture and forest catchments in the upper Mara River basin, Kenya; [¥]identifies forest streams; sampled only during the [#]wet and *dry seasons, respectively.

Site	Wet season				Dry season			
	GPP (gO$_2$ m^{-2} day^{-1})	ER (gO$_2$ m^{-2} day^{-1})	NEP (gO$_2$ m^{-2} day^{-1})	GPP/ER	GPP (gO$_2$ m^{-2} day^{-1})	ER (gO$_2$ m^{-2} day^{-1})	NEP (gO$_2$ m^{-2} day^{-1})	GPP/ER
Issey I[¥]	0.52	-4.83	-4.31	0.11	1.63	-5.66	-4.02	0.29
Issey II[¥]	0.29	-2.70	-2.41	0.11	1.34	-7.13	-5.78	0.19
Bondet[¥]	0.18	-2.02	-1.84	0.09	0.14	-8.09	-7.95	0.02
Mosoriot[¥]	0.13	-1.04	-0.91	0.13	1.47	-7.72	-6.25	0.19
Sara[¥]	0.70	-4.12	-3.43	0.17	1.66	-4.52	-2.86	0.37
Issey IV[#]	1.33	-8.02	-6.69	0.17	-	-	-	
Masai*	-	-	-		3.59	-7.01	-3.42	0.51
Kips	2.76	-6.09	-3.33	0.45	11.53	-31.12	-19.59	0.37
Bomet	14.39	-49.17	-34.78	0.29	15.34	-56.87	-41.53	0.27
Kaki	1.47	-2.88	-1.41	0.51	7.20	-7.86	-0.66	0.92
TNK	0.71	-3.08	-2.37	0.23	3.82	-4.52	-0.70	0.85

in AGR streams the range was from -41.53 g O$_2$ m^{-2} day^{-1} to -0.66 g O$_2$ m^{-2} day^{-1} (P/R range: 0.27–0.92). At all streams GPP rates were lower than ER rates, and using the 0.5 as the thresholds between autotrophy (P/R > 0.5) and heterotrophy (Meyer, 1989), all forest streams were heterotrophic during both the dry and wet seasons. In comparison, one and three AGR streams were autotrophic during the wet and dry seasons, respectively.

Stepwise multiple regression models identified predictors of ecosystem metabolism that are linked to catchment and riparian land use (Table 5.6). Ln TDN, proportion of catchment area under agriculture (arcsine % AGR) and stream size (streamsize-PC1 and streamsize-PC2 for ER) were linked to GPP, ER and NEP during the dry season. During the wet season, canopy cover, and stream size (wetted width) were linked to GPP rates, in addition to nutrients (Ln TDN) and agricultural land use (arcsine % AGR). Nutrient loadings (Ln TDN) and canopy cover were also linked to ER rates and NEP.

3.4 Linkage between metabolism and DOM composition

There were significant relationships between GPP and ER rates and DOM composition (Figure 5.6). GPP and ER rates during the dry season were positively related to DOM-PC1 (simple linear regression, $R^2 = 0.61$, $p = 0.005$, $n = 10$ and $R^2 = 0.45$, $p = 0.020$, $n = 10$, respectively) and negatively related to DOM-PC2 ($R^2 = 0.73$, $p = 0.001$, $n = 10$ and $R^2 = 0.62$, $p = 0.004$, $n = 10$). DOM-PC1 was associated with low molecular weight, freshly and autocthonously produced DOM, as indicated by its strong positive correlation with the fluorescence index (FI, $r = 0.77$, $p = 0.007$). DOM-PC2 was positively correlated with higher molecular weight DOM and aromaticity (SUVA$_{254}$, $r = 0.71$, $p = 0.019$) and negatively correlated with the freshness index (β:α, $r = -0.84$, $p < 0.0001$). Consequently, GPP and ER rates were related with FI and β:α during the dry season (Fig. 7b,d) and with FI during the wet season (Fig. 7d,h). In addition, GPP and ER rates were related to DOM-PC2 during the wet season ($R^2 = 0.64$, $p = 0.003$, $n = 10$ and $R^2 = 0.58$, $p = 0.004$, $n = 10$, respectively). DOM-PC2 was positively related to FI ($r = 0.86$, $p < 0.0001$) indicating that this PC-axis was associated with freshly and autochthonously produced DOM.

4. Discussion

4.1 Effects of land use change on DOM composition

I report notable shifts in DOM composition and ecosystem metabolism as a result of catchment land use change from forestry to agriculture. The land use influence was also expressed through water physico-chemistry with major ions, suspended solids and dissolved nutrients increasing in agriculture streams (Figure 3ab). FOR streams were associated with higher molecular weight and terrestrially derived DOM, while AGR streams were associated with fresher, autochthonously produced and low molecular weight DOM (Figure 3cd).

Land use driven shifts in DOM composition were also captured by recently produced (β:α), autochthonously produced DOM (BIX) and the slope ratio (S_R) (Table 5.1). In this study, β:α indicates increased autochthonous production of DOM in AGR and MIX streams (Wilson and Xenopoulos, 2009). Increased autochthonous DOM production in AGR and MIX streams was further supported by BIX whereby values >0.8 in MIX and AGR streams were indicative of predominantly fresher and autochthonous produced DOM (Huguet et al., 2009). Higher S_R values in AGR streams were evidence of photodegradation as a result of exposure to sunlight as a result of open canopy, and this lowers the molecular weight of DOM (Helms et al., 2008). Predominance of low molecular weight DOM in AGR streams was further supported by low a_{254}/a_{410} values (Baker et al., 2008).

However, despite a significant influence of catchment land use on the emergent patterns of DOM composition, reach-scale influences were also noted with sites not completely separating according to land use along PCA-axes (Figure 3c). In-stream activities by livestock and people in agriculture streams were patchy and led to losses of canopy cover and wider channels (Minaya et al., 2013; Masese et al., 2014a, b). Similar values of molecular weight (E_2:E_3) and short wavelength slope ($S_{275-295}$) at FOR, MIX and AGR streams (Table 5.1) captured the effects of pockets of canopy cover along streams which were sources of allochthonous litter input. On the contrary, areas with open canopy and wider channels exposed streamwater to photodegradation. Additionally, the low FI values (1.3-1.4) were indicative of a predominance of terrestrially derived and structurally complex DOM (Fellman et al., 2010; Huguet et al., 2009) in both AGR and FOR streams. Evidence of structurally complex DOM in AGR streams was further supported by higher $SUVA_{254}$ values. In the study area livestock watering in streams is a

Table 5.6. Results of step-wise multiple linear regression for controls of rates of gross primary production (GPP), ecosystem respiration (ER) and net ecosystem production (NEP) ($n = 10$ for each analysis).*

Dependent variable	Independent variable	Parameter estimate (SE)	R^2	P – value
Dry season				
GPP	Intercept	-2.743 (0.765)		
	Ln TDN	2.827 (0.425)	0.850	<0.001
	Arcsine % AGR	2.716 (0.764)	0.058	0.012
	Streamsize-PC1	-0.723 (0.264)	0.051	0.034
	Full model		0.959	<0.001
ER	Intercept	-10.757 (0.871)		
	Ln TDN	15.374 (0.478)	0.961	<0.001
	Arcsine % AGR	-2.699 (0.654)	0.013	0.009
	Streamsize-PC1	-0.721 (0.227)	0.005	0.025
	Streamsize-PC2	4.876 (0.682)	0.018	<0.001
	Full model		0.997	<0.002
NEP	Intercept	8.661 (0.816)		
	Ln TDN	12.549 (0.458)	0.889	<0.001
	Arcsine % AGR	4.877 (0.609)	0.057	<0.001
	Strmsize-PC2	-5.511 (0.648)	0.049	<0.001
	Full model		0.995	<0.001
Wet season				
GPP	Intercept	22.145 (3.630)		
	Arcsine % canopy cover	-3.946 (0.780)	0.671	0.004
	Ln TDN	3.316 (0.582)	0.154	0.002
	Arcsine % AGR	2.832 (1.163)	0.092	0.059
	Ln wetted width	0.978 (0.439)	0.042	0.076
	Full model		0.959	0.001
ER	Intercept	83.661 (13.419)		
	Arcsine % canopy cover	-17.425 (3.196)	0.596	<0.001
	Ln TDN	7.010 (2.351)	0.226	0.020
	Full model		0.822	0.002
NEP	Intercept	-58.912 (9.598)		
	Arcsine % canopy cover	12.062 (2.286)	0.559	0.001
	Ln TDN	5.275 (1.681)	0.258	0.016
	Full model		0.817	0.003

*Note: Considered variables include streamsize-PC1-4, water-quality (WQ)-PC1-4, Ln TDN, Ln DOC concentration, Ln FBOM, Ln CPOM, Ln C:N ratio, arcsine % canopy cover, arcsine % AGR land use, arcsine % POM in TSS, Ln NO_3-N and Ln NH_4^+. Criterion for entry into the model was $p = 0.05$, except for GPP during the wet season where results for a lenient entry criterion ($p = 0.1$) are also given.

common occurrence and the footpaths created by their hoof action can be major sources of inert pools of DOM, in addition to farmlands, when it rains. Land use effects on TSS were stronger than on POM (Table 5.1), suggesting increased soil erosion in the agriculture-dominated catchments (Foley et al., 2005; Quinton et al., 2010).

4.2 Stream size and longitudinal variability in DOM composition

The proportion of recently ($\beta:\alpha$) and autochthonous produced DOM (BIX) increased with river distance and stream size (Figure 5.4). This is consistent with increasing primary production as streams widen and canopy cover reduces (Vannote et al., 1980). The longitudinal trends observed for total absorption (200-600 nm) agree with shifts in absorbance as large molecules of DOM originating from forest is

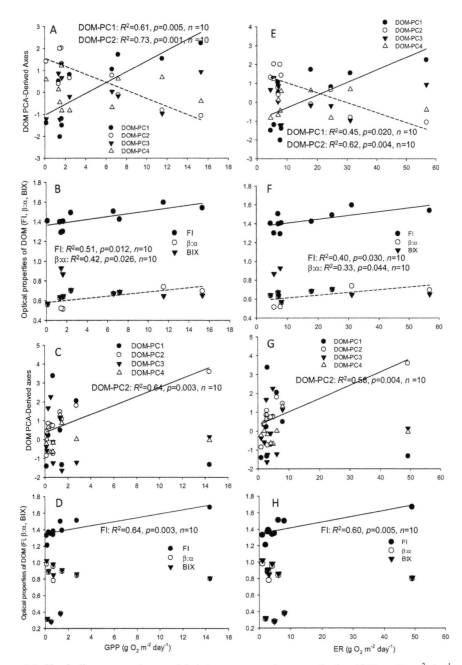

Figure 5.6. Simple linear regressions models between gross primary production (GPP, g O_2 m^{-2} day^{-1}) and ecosystem respiration (ER, g O_2 m^{-2} day^{-1}) rates and dissolved organic matter (DOM) composition by season. The optical properties considered were the fluorescence index (FI), freshness index (β:α) and the autochthonous biological index (BIX) and the first four PC-axes derived from PCA ordinations of DOM composition data. During the dry season (a, b, e, f) GPP rates were related to (a) DOM –PC1 and DOM-PC2 and (b) FI and β:α, and ER rates were related to the same variables (e) DOM –PC1 and DOM-PC2 and (f) FI and β:α. During the wet season (c, d, g, h) GPP rates were related to (c) DOM-PC2 and (d) FI, and ER rates were related to the same variable (g) DOM-PC1 and (h) FI.

115

photochemically broken down along the river continuum (Helms et al., 2008). However, because of changes in stream characteristics as a result of land use change, e.g., loss of canopy cover and widening of channels, longitudinal patterns resemble patterns across a land use gradient (McTammany et al., 2007). Consequently, longitudinal patterns observed for some variables were likely confounded by land use change since most of the largest streams fall under the mixed land use category (Figure 5.1 and 5.4).

Evidence for the influence of land use on longitudinal patterns was observed for aromaticity ($SUVA_{254}$) and short wavelength slope (S_R) which reduced with river distance (Figure 5.4). It would be expected that S_R would increase in MIX streams because of increased photodegradation as streams widen (Helms et al., 2008), and that $SUVA_{254}$ would display higher values in FOR streams where aromatic terrestrial sources dominate as compared with AGR and MIX streams. For $SUVA_{254}$ a number of factors can explain the higher values observed in AGR streams compared with FOR streams. Firstly, sampling was done towards the end of the rainy season, and it has been observed that DOC concentrations, which would have tracked $SUVA_{254}$ in this case, follow hysteresis with a nearly linear increase during the rising limb of the hydrograph, followed by a gradual drop during peak discharge and falling limb (Sanderman et al. 2009; Lambert et al. 2011; Bouillon et al., 2012). It is therefore likely that at the time of sampling, forest soils had been flushed effectively reducing the levels of aromatic and higher molecular weight DOC. On the other hand, land tillage has been reported to mobilize inert pools of DOM increasing the input of higher molecular weight DOM in streamwater (Ogle et al., 2005; Graeber et al., 2012). Evidence of increased soil erosion in AGR streams is captured by significantly higher levels of TSS and lower POM and proportions of carbon in TSS (Table 5.1). Secondly, aromaticity in AGR streams could have also been due to selective utilization of labile DOM by microbes (Ortega-Retuerta et al., 2009). Indeed, GPP and ER rates in these streams are coupled (details below) suggesting that much of the labile autochthonous produced DOM is respired (Townsend et al. 2011; Griffiths et al. 2013), and this selective utilization could increase the proportion of recalcitrant allochthonous DOM in AGR streams. However, priming of recalcitrant DOM pools by labile DOM has been seen to foster its metabolism, further coupling GPP and ER rates and driving streams more into heterotrophy (Guenet et al., 2010; Townsend et al., 2011). Lastly, it has been demonstrated that aromatic compounds deriving from lignin degradation dominate the DOM fraction absorbing at 254 nm (Kalbitz et al., 2003). Most of the AGR streams are lined with *Eucalyptus* spp. whose leaves are known to have high concentrations of lignin and phenolic compounds, and their leachates could be contributing to higher $SUVA_{254}$ values. Evidently, the behavior of DOM pools in streams in this study was complex and subject to a number of controls whose effects are difficult to distinguish.

4.3 Ecosystem metabolism

To my knowledge, this is the first study to directly measure GPP and ER rates in African tropical streams. However, the results of this study are comparable to studies in both tropical and temperate

Table 5.7. Comparison of GPP, ER and NEP rates across 1st-3rd order headwater streams from different regions around the world and dry season results of the present study. For all streams the open-channel method (Marzolf et al. 1994, Young and Huryn 1998) was used to estimate metabolic rates.*

Stream	Region, climate	GPP (gO_2 m^{-2} day^{-1})	ER (gO_2 m^{-2} day^{-1})	NEP (gO_2 m^{-2} day^{-1})	Land use/ comments	Reference
Ag North	Kansas, temperate	7.03	6.94	0.09	Agriculture	Riley and Dodds, 2013
N04D	Kansas, temperate	10.17	6.08	4.09	Bison grazing	Riley and Dodds, 2013
Campus Creek	Kansas, temperate	1.37	2.43	-1.06	Urban	Riley and Dodds, 2013
Natalie's Creek	Kansas, temperate	0.78	3.45	-2.67	Cattle grazing	Riley and Dodds, 2013
Swine Creek	Kansas, temperate	4.94	3.77	1.17	Mixed	Riley and Dodds, 2013
Shane Creek	Kansas, temperate	3.45	5.3	-1.85	Prairie	Riley and Dodds, 2013
Hugh White Creek	North Carolina, temperate	0.21	10.1	-9.89	Forest	Mulholland et al., 1997
Walker Branch	Tennessee, temperate	0.32	4.12	-3.8	Forest	Mulholland et al., 1997
Walker Branch	Tennessee, temperate	1.4	4.0	-2.6	Forest	Roberts et al., 2007
Sycamore Creek	Arizona, temperate	15	8.3	6.7	-	Mulholland et al., 2001
Ditch Creek	Wyoming, temperate	1.94	6.45	-4.51	Little disturbance	Hall and Tank, 2003
Pilgrim Creek channel 2	Wyoming, temperate	0.13	1.59	-1.46	Little disturbance	Hall and Tank, 2003
Spread Creek	Wyoming, temperate	3.11	8.37	-5.26	Little disturbance	Hall and Tank, 2003
Glade Creek	Wyoming, temperate	1.08	13.3	-12.22	Little disturbance	Hall and Tank, 2003
Bailey Creek	Wyoming, temperate	1.04	2.02	-0.98	Little disturbance	Hall and Tank, 2003
S1	Iceland, temperate	12.99	25.46	-12.47	Grassland dominated	Rasmussen et al., 2011
S2	Iceland, temperate	9.64	14.72	-5.08	Grassland dominated	Rasmussen et al., 2011
S3	Iceland, temperate	2.37	9.44	-7.07	Grassland dominated	Rasmussen et al., 2011
S4	Iceland, temperate	0.66	5.55	-4.89	Grassland dominated	Rasmussen et al., 2011
Bears Creek	Manitoba, temperate	3.21	9.62	-6.41	Mixed	Yates et al., 2012
N. Shanno Creek	Manitoba, temperate	10.08	14.94	-4.86	Mixed	Yates et al., 2012
Deadhorse Creek	Manitoba, temperate	11.38	10.55	0.83	Mixed	Yates et al., 2012
Q. Concepción	Peruvian Amazon, tropical	0.07	5.17	-5.1	2nd growth forest	Bott and Newbold, 2013
Q. Abejitas	Peruvian Amazon, tropical	0.19	15.02	-14.83	Pasture	Bott and Newbold, 2013
Q. Tambopata	Peruvian Amazon, tropical	0.08	1.88	-1.8	Forest	Bott and Newbold, 2013
Tai Po Kau	Hong Kong, China, tropical	0.09	-	-	Forest, riffle	Dudgeon, 1999
Tai Po Kau	Hong Kong, China, tropical	0.22	-	-	Forest, Pool	Dudgeon, 1999
Bisley	Puerto Rico, tropical	0.19	2.44	-2.25	Natural forest	Ortiz-Zayas et al., 2005
Puente Roto	Puerto Rico, tropical	1.77	4.07	-2.3	Natural forest	Ortiz-Zayas et al., 2005
La Vega	Puerto Rico, tropical	1.15	4.96	-3.81	Natural forest	Ortiz-Zayas et al., 2005
Pristine stream 1	Brazil, tropical	<0.1	8	-7.9	Natural forest	Gücker et al., 2009
Pristine stream 2	Brazil, tropical	0.3	6.2	-5.9	Natural forest	Gücker et al., 2009
Pristine stream 3	Brazil, tropical	0.2	5.6	-5.4	Natural forest	Gücker et al., 2009
Agriculture stream 1	Brazil, tropical	0.4	4.3	-3.9	Agriculture	Gücker et al., 2009
Agriculture stream 2	Brazil, tropical	0.8	3.2	-2.4	Agriculture	Gücker et al., 2009
Agriculture stream 3	Brazil, tropical	0.8	2.2	-1.4	Agriculture	Gücker et al., 2009
Many streams	Australia, tropical	6.1(<0.01, 29.9)*	6(0.1, 23.4)*	-0.1(-11.4, 18.4)*	Mixed land uses	Fellows et a., 2006
Mitchell River[#]	Australia, tropical	2.12	4.47	-2.35	Forest	Hunt et al., 2012

*Note: values presented include the mean and the min and max values in brackets. [#]the river is more than 3rd order.

streams that have used the open-system method (Table 5.7). In forest streams, the range of GPP rates during the dry and wet seasons (0.13-1.66 g O_2 m^{-2} day^{-1}) were similar to ranges reported for forest streams in Puerto Rico (GPP, 0.19-1.77 g O_2 m^{-2} day^{-1}; Ortiz-Zayas et al., 2005) and Pennsylvania (0.62-1.85 g O_2 m^{-2} day^{-1}; Bott et al., 2006a). For agriculture streams, the range of GPP rates was much wider (0.71-15.34 g O_2 m^{-2} day^{-1}) but fall within ranges reported for a number of steams draining mixed land uses in tropical Australia (<0.01-29.9 g O_2 m^{-2} day^{-1}; Fellows et al., 2006). The ranges are also similar to rates in a number of streams in different biomes and land uses (<0.1-15 g O_2 m^{-2} day^{-1}; Mulholland et al., 2001) and in those draining mixed land uses in Kansas (0.78-10.17 g O_2 m^{-2} day^{-1}; Riley and Dodds, 2013).

Comparatively, the range of ER rates in forest streams (-1.04 to -8.09 g O_2 m^{-2} day^{-1}) were similar to values in forest streams in Puerto Rico (-2.44 to -4.96 g O_2 m^{-2} day^{-1}; Ortiz-Zayas et al., 2005), New Zealand (-1 to -8 g O_2 m^{-2} day^{-1}; Young and Huryn, 1999), and a number of streams draining mixed land uses in New York (-1.39 to -8.30 g O_2 m^{-2} day^{-1}; Bott et al., 2006b) and Wyoming (-2.02 to -8.37 g O_2 m^{-2} day^{-1}; Hall and Tank, 2003). For agriculture streams, the range for both the dry and wet seasons (-2.88 to -56.87 g O_2 m^{-2} day^{-1}) represent one of the highest ranges in the literature. However, the highest ER rate was measured in a stream (Bomet) that is influenced by nutrients and organic waste inputs from a town nearby. Bomet stream also experiences reduced dry season flows (Table 5.3) and has a flushy flow during the wet season. Excluding Bomet stream, ER rates in agriculture streams (range -2.88 to -34.78 g O_2 m^{-2} day^{-1}) are similar to ranges obtained during continuous one year-long measurements in an agricultural stream in mid-western United States (-0.9 to -34.8 g O_2 m^{-2} day^{-1}, Griffiths et al., 2013), and first- and second-order streams in a prairie watershed in Illinois (-6.2 to -34.0 g O_2 m^{-2} day^{-1}; Wiley et al. 1990). The ER rates for three out of the five agriculture streams also fall within ranges in streams draining mixed land uses in tropical Australia (-0.1 to -23.4 g O_2 m^{-2} day^{-1}; Fellows, et al., 2006).

Compared with forest streams, agriculture streams were more metabolically active with higher rates of GPP and ER during both the dry and wet seasons. The lower nutrient levels and high canopy cover in forest streams can explain the comparatively lower rates in FOR streams. On the contrary, canopy cover in agriculture streams was much reduced, $\leq$ 50% overall and light availability and higher nutrient levels can explain the higher GPP rates (Roberts et al., 2007; Griffiths et al., 2013).

GPP had a significant positive correlation with ER (r = 0.85, p < 0.001, n = 20) suggesting a tight coupling of the two processes. The occurrence and strength of the relationship between ER and GPP differs among stream systems. A study on streams in New York also showed a positive relationship between GPP and ER (Bott et al., 2006b) as was the case with Wiley et al. (1990) and Bunn et al. (1999). However, some studies have found weak or non-existent relationships (Mulholland et al., 2001). In this study, ER was consistently much higher than GPP, indicating that these streams are heterotrophic, as would be expected for headwater streams (Vannote et al. 1980). However, GPP and

ER rates were higher at AGR streams suggesting a shift towards autotrophy, especially during the dry season. Shifts from heterotrophy to autotrophy following land use change are consistent with a number of studies in tropical and temperate biomes (Bunn et al., 1999; Young et al., 2008), although in some cases similar findings have not been reported (Young and Huryn 1999; Bernot et al. 2010). Moreover, the threshold for autotrophy and heterotrophy is itself controversial. While a value of P/R equal to 1 has been used as the boundary, Meyer (1989) suggested using a P /R < 0.5 for streams supported predominantly by allochthonous inputs. The explanation is that if a stream has a P/R equal to 1, a small input of allochthonous material would drive it into heterotrophy (Rosenfeld and Mackay 1987). Using P/R equal to 0.5 as the boundary in this study, most of the agriculture streams were predominantly fuelled by autochthonous resources, less so during the wet season, whereas all forest streams were predominantly fuelled by allochthonous resources. As an integrated assessment of total metabolic activity of the ecosystem in the study reaches, the more negative NEP values for agriculture streams would suggest that they were more heterotrophic. However, this can be interpreted also to mean that agriculture streams processed significantly greater amounts of organic matter per unit channel length than forest streams. Indeed, agriculture streams were recipients of both animal and human wastes that included excreta, sediments and organic wastes. Higher ER rates in agriculture streams can also be linked to higher nutrient levels and temperature.

Measuring ecosystem metabolism in streams is challenging because of diel variability in reaeration which makes extrapolation of daylight (most re-aeration experiments are done during the day) estimates for 24 hour period sometimes unrealistic. This is especially so during the wet season when re-aeration changes with rain events. In this study, I experienced a series of heavy storms during the wet season that likely affected estimates of re-aeration. However, during the dry season I was able to measure and calculate re-aeration using propane evasion and the energy dissipation methods, respectively. The dry and wet season calculated (EDM) reaeration rates were significantly correlated ($r = 89$, $p<0.001$, $n = 10$). The measured and calculated reaeration (EDM) rates during the dry season were also significantly correlated ($r = 84$, $p = 0.002$, $n = 10$). Similarly, Riley and Dodds (2013) found a significant relationship (Kendall's τ correlation analysis $= 0.73$, $p = 0.039$) between measured and calculated (EDM) k in Kansan streams. These relationships support my estimates of both GPP and ER rates which are largely comparable with previous studies in the tropics (Table 5.7).

4.4 Factors controlling ecosystem metabolism

Significant spatial variability in GPP and ER rates were noted among streams and linked to both local and reach-scale influences. TDN concentration, stream size (channel width), canopy cover and the proportion of agricultural land use in the catchment (arcsine % AGR) were linked with GPP, ER and NEP during the dry season (Table 5.6). These variables are good proxies for both catchment-scale (proportion of agricultural land use) and reach-scale (nutrient loading and widening of stream channels and loss of canopy cover) influences in the study area. The stream size effect was related to the combined effects of reducing canopy cover and increasing primary production along the riverine continuum (Vannote et al., 1980). The lower canopy levels and wider channels implies higher levels of photosynthetically active radiation, which I did not measure but is a key factor controlling metabolism in streams and rivers (Mulholland et al., 2001; McTammany et al., 2007). Higher TDN concentrations in AGR streams can be attributed to runoff from farmlands (farmers use nitrogenous fertilizers for topdressing) and inputs by livestock when watering in streams, as well as leakages from toilets in human settlements. Consequently, at Bomet and Kips streams inputs of nutrients from Bomet town (20,000 people without sewerage services) and by livestock, respectively, likely contributed to the high GPP and ER rates. Evidence of input of nitrates from human settlements and animal excreta into agriculture streams in these streams has been supported by $\delta^{15}N$ values of basal resources (including filamentous algae and periphyton) and invertebrates (Masese et al., 2015).

Seasonality was an important factor and through its control on canopy cover and discharge in both forest and agriculture streams (Table 5.3), longitudinal connectivity and terrestrial-aquatic hydrologic-linkages, dry- and wet-season differences in GPP and ER rates were observed. Higher rates of GPP and ER were recorded during the dry compared with the wet season. Seasonal changes in discharge can cause storms that trigger scour and deposition in streams with an influence on metabolism rates (Roberts et al., 2007; Griffiths et al., 2013). During the wet season increased levels of turbidity likely smothered streambeds and limited light availability for primary production. Even though conducted during short periods in the dry and wet seasons, the findings of this study show that GPP and ER rates can change seasonally in response to wet-dry conditions caused by rainfall variability, as opposed to the temperature- and light-driven seasonal changes in temperate streams. However, yearlong continuous studies are needed in these streams to capture the whole range of seasonal variability and determine the longterm controls of DOM composition and metabolism in these streams (Roberts et al., 2007; Jaffe et al., 2008).

4.5 Linkage between DOM composition and ecosystem metabolism

Land use driven shifts s in DOM composition can influence its availability for metabolism and, reciprocally, ecosystem metabolism (GPP and ER) can affect DOM composition via consumption or production through primary production (Cammack et al., 2004; Halbedel et al., 2013; Barrón et al., 2014). I hypothesized that these dependencies should lead to pronounced relationships among measures of ecosystem metabolism and DOM composition in streams draining a land use gradient (see Halbdelet al., 2013). Optical properties of DOM largely agreed with direct measures of ecosystem metabolism as demonstrated by significant relationships between DOM indices and GPP and ER rates (Figure 5.6). During the dry and wet seasons, GPP and ER rates were positively related with the fluorescence index and PCA-derived DOM-axes associated with freshly and autochthonously produced DOM (OM-PC1) during the dry season and DOM-PC2 during the wet season. The higher GPP rates in AGR streams, which were associated with fresher and autochthonously derived DOM of lower molecular weight, suggest a linkage between DOM composition and ecosystem metabolism in these streams. It is likely that the autochthonously produced DOM increased metabolism rates in AGR streams because of its lability and bioavailability. Since GPP and ER rates in these streams were tightly coupled, it is also likely that much of the carbon produced was respired rather than incorporated into biomass (Bunn et al., 1999; Griffiths et al., 2013). Additionally, changes in the DOM matrix driven by photochemical processes can boost the respiration of microbial communities (Anesio et al., 2000). Therefore, the higher rates of ER in AGR streams could have also been driven by photodegradation of allochthonous DOM, as evidenced by the higher S_R values (Table 5.1).

In comparison with AGR streams, FOR streams recorded lower rates of GPP, and hence were more heterotrophic and mainly fuelled by allochthonous resources (Vannote et al., 1980). This is also confirmed by the predominance of terrestrially derived and higher molecular weight DOM. The high canopy cover and low nutrient levels in FOR streams did not support primary production, in comparison with AGR streams that were wider (low canopy cover) and with higher nutrient levels.

4.6 The land use gradient and downstream implications

The findings of this study have important implications for the functioning of the Mara River as a system. In addition to the results of this study, previous works in streams and rivers have shown that land use change and increased inputs of bioavailable carbon and nitrogen can have strong influences on DOM composition and concentration and productivity (Bernot et al., 2010; Halbedel et al., 2013; Cawley et al., 2014). The Mara River basin has witnessed extensive deforestation in its upper reaches and conversion of savanna grasslands into settlements and farmlands in the middle reaches (Mati et al., 2008). Moreover, indigenous riparian vegetation is being replaced with exotic species, changing organic matter input regimes, its quality and the structure of benthic invertebrate communities (Masese et al., 2014a, b). Increased microbial processes in agriculture streams have also been reported with implications on nutrient cycling, organic matter availability and energy transfer to higher trophic levels (Masese et al., 2014b, 2015). These upstream changes have implications on the Mara River and the provision of valuable services such as water quality, support for wildlife, fisheries and recreation. The findings of this study and the consequences they signify for the Mara River reinforces a need for improved understanding of land use-driven and longitudinal changes in organic matter quality, processing and export since these are critical for predicting downstream alterations of ecosystem functions and provision of services (Seitzinger and Sanders 1997; Petrone et al. 2009). Given that forest conversion is increasing across many watersheds globally (Foley et al. 2005), changes in the processing and composition of dissolved carbon in streams and rivers are likely to increase in the future.

Conclusions

This study identified effects of land use on the sources and composition of dissolved organic matter (DOM) and ecosystem metabolism in upland streams in the Mara River, Kenya. I was able to distinguish between different and spatially varying properties of DOM in forest and agriculture streams. I used a combination of fluorescence and absorbance indices of DOM, and whole-stream metabolism as a new integrated approach for assessing the functioning of streams across a land use gradient. Compared with agricultural streams, forested streams exported structurally complex and high molecular weight DOM, a difference that was significant during both the dry and wet seasons. However, aromaticity of DOM was high at all sites regardless of land use, suggesting a predominance of ubiquitous terrestrial source of DOM even in agriculture streams. In agriculture streams aromaticity could be linked to erosion of the upper soil layers from farmlands and subsequent transport into streams by runoff. During the rains, agriculture streams were much turbid compared with forest streams. Measures of ecosystem metabolism responded to the land use change with higher rates of GPP and ER in agriculture streams. DOM composition seemed to track this variability; forest streams were heterotrophic and hence transported terrestrially derived DOM, whereas agriculture streams were more autotrophic and transported a mixture of DOM that was dominated by fresh and autochthonously produced DOM. Seasonality was important and through its control on canopy cover (Table 5.3) and hydrologic linkages, dry- and wet-season differences in GPP and ER, and in extension the composition of DOM, were observed. Analysis of DOM composition on the longitudinal gradient of streams captured shifts from predominantly allochthonous to predominantly autochthonous inputs which agree with the river continuum concept (Vannote et al., 1980). Stream size mainly influenced the molecular size of DOM through photodegradation as streams became wider and insolation increased. However, the longitudinal gradients also mimicked the land use gradient making is difficult to reconcile patterns in some properties of DOM, such as reducing $SUVA_{254}$ and S_R with increasing stream size.

These findings help elevate the role of land use and specifically agriculture, as a source of structurally altered DOM to surface waters. The results also demonstrate a link between DOM composition and ecosystem metabolism in headwater streams influenced by forest and agriculture land use. Potential differences in catchment characteristics, and for my case land use history and usage and the techniques - plough and hoe vs intensive mechanization- need to be addressed in future studies to better constrain the effect of different agricultural practices and intensities on DOM quantity and composition in aquatic ecosystems. Measures of ecosystem metabolism in these streams have helped compliment the role of tropical streams and rivers as important components of the global carbon cycle and adds to the growing understanding of the effects of agriculture on ecosystem functioning.

Chapter

6

Partitioning the relative importance of different sources of energy for consumers on the longitudinal gradient of the Mara River, Kenya

Publication to be based on this chapter:
Masese FO, Abrantes KG, Gettel GM, Bouillon S, Irvine K, McClain ME. (2015). Large herbivores as vectors of terrestrial subsidies for riverine food webs. *Ecosystems,* in press.

Abstract

The African tropical savanna has undergone rapid land-use change, accompanied by a dramatic reduction in native large mammalian herbivore populations. Consequences of these developments for energetic terrestrial-aquatic linkages are largely unknown. In this study the importance of herbivore-mediated subsidies for consumers were determined in the Mara River, Kenya by quantifying spatial and temporal patterns of carbon flow using natural abundances of stable carbon ($\delta^{13}C$) and nitrogen ($\delta^{15}N$) isotopes. Potential primary producers (terrestrial C3 and C4 producers and periphyton) and consumers (invertebrates and fish) were collected from sites along the river that represent different effects of human and herbivore (livestock and wildlife) influence. Sampling was conducted during the dry and wet seasons to represent a range of contrasting flow conditions. Stable Isotope Analysis in R (SIAR) Bayesian mixing models was used to partition terrestrial and algal sources of organic carbon supporting consumer trophic groups. The relative importance of organic carbon sources differed among sites and with season. Overall periphyton dominated contributions to consumers during the dry season. During the wet season the importance of terrestrial-derived carbon for consumers was higher with the importance of C3 producers declining with distance from the forested headwaters as the importance of C4 producers increased in river reaches receiving livestock and hippo inputs. This study highlights the importance of large mammalian herbivores on the functioning of riverine ecosystems and the implications of their loss from savanna landscapes that currently harbour remnant populations. While the importance of C4 terrestrial carbon in most river systems has been reported to be negligible, the evidence from this study suggest its contribution to streams, mediated by livestock, is widespread.

1. Introduction

The importance of different sources of organic carbon to riverine food webs has been postulated to vary longitudinally along the river continuum (Vannote et al., 1980; Junk et al., 1989; Thorp and Delong, 2002). Transfers of terrestrial organic matter and nutrients can provide important subsidies to receiving aquatic ecosystems, enhancing primary and secondary production (Polis et al., 1997; Paetzold et al., 2007). While much of organic matter and nutrient fluxes into streams occurs through direct litterfall from riparian vegetation and hydrologic transport through surface and sub-surface flowpaths, the movement of herbivores can actively transfer organic matter and nutrients into river through defecation and urination in the water (Naiman and Rodgers, 1997; Grey and Harper, 2002; Bond et al., 2012). However, large populations of herbivorous mammals that were once key features of many landscapes in temperate and tropical biomes have been decimated by human actions and replaced to some extent by domesticated cattle (Prins, 2000; Ogutu et al., 2011; Wardle et al., 2011). The effects of this occurrence on energetic terrestrial-aquatic food web linkages are largely unknown.

Trophic resources are dynamic in space and time depending on prevailing environmental conditions, and this dynamism is propagated to the structure of food webs (Woodward and Hildrew, 2002; de

Ruiter et al., 2005). For instance, flow variation in rivers influences ecosystem size, organic matter flux, light and nutrient availability for primary production (Power et al., 1995; Tank et al., 2010). Subsidy pathways can also change seasonally because of changes in connectivity, flowpaths and transport vectors (Paetzold et al., 2007; Wipfli and Baxter, 2010). For instance, during the dry season, higher metabolic activity means that animals need to visit watering points more frequently (Bond et al., 2012). In savanna landscapes, water availability is a strong determinant of herbivore distributions with many herbivores congregating near watering points (du Toit et al., 1990; Ogutu et al., 2010). The effects of transfers also depend on the quality of the subsidy relative to local resources. Algal carbon contributes significantly to metazoan biomass in mid-sized and large rivers despite forming a small proportion of available food resources, with detritus from vascular plants playing only a minor role (Lewis et al., 2001; Thorp and Delong, 2002). Studies have also shown that because of its poorer quality C4 grasses contribute minimally to food webs compared with C3 vegetation (Clapcott and Bunn, 1997; Roach, 2013), and basal production sources supporting consumers can change seasonally (Huryn et al., 2001; Zeug and Winemiller, 2008).

Stable isotope analysis (SIA) provides a time-integrated measure of carbon flow in food webs (Fry and Sherr, 1989; Post, 2002). The ratio of ^{13}C to ^{12}C isotopes (expressed as $\delta^{13}C$ values) is used to partition different organic carbon sources and to infer energy flow through food webs because of the small fractionation (0–1%) from food source to consumer (Fry and Sherr, 1989; McCutchan et al., 2003). Alternatively, the ratio of ^{15}N to ^{14}N isotopes (expressed as $\delta^{15}N$ values) is used to infer the trophic position of a consumer (Post, 2002). The use of two or more isotopes strengthens the discrimination between potential food sources, especially in cases where sources overlap in one of the isotopes (McCutchan et al., 2003). However, the isotopic composition of potentially important aquatic primary producers such as periphyton can vary spatially and temporally (Finlay, 2004; Hadwen et al., 2010b). Therefore, sampling across seasons to capture potential variability in primary producer isotopic composition is important for estimating of food web dynamics in aquatic ecosystems.

East African rivers display highly seasonal flow regimes with well defined wet and dry seasons (McClain et al., 2014), leading to annual cycles in habitat and nutrient availability and productivity (Marwick et al., 2014a). The degree of terrestrial-aquatic linkage in the supply of nutrients and carbon is influenced by seasonal hydrologic changes reported to be greatest in deforested and grazing areas during the wet season (Augustine et al., 2003; Dutton 2012; Defersha and Melesse, 2012). The input of animal-mediated subsidies is dependent on animal behaviour and population densities (Plate 6.1) (Bond et al., 2012; 2014). Human activities on catchments and riparian corridors influence the distribution of vegetation and mammalian herbivores thereby directly influencing the functioning of rivers. For example, deforestation mobilises nutrients and suspended matter, increasing turbidity and limiting primary production during the wet season (Huryn et al., 2001; Mead and Wiegner, 2010). The loss of large populations of herbivores through conversion of their grazing lands to agriculture likely reduces the supply of terrestrial organic matter and nutrients to rivers.

The Mara River (Kenya, Tanzania) traverses a landscape gradient that presents a unique case for studying the influence of both human and animal populations on terrestrial-aquatic food web linkages. Upper reaches are forested, with a transition into mixed small-scale and large-scale agriculture and human settlements at the foot of the Mau Escarpment (Mati et al., 2008). Livestock and land use change have been linked to shifts in the characteristics of organic matter and nutrients (Masese et al., 2014a), with likely influences on the structure and functioning of the river (Plate 6.1).

The aim of this study is to investigate energy sources for consumers in river reaches strongly influenced by terrestrial subsidies and how these subsidies interact with seasonal discharge (seasonality) to influence ecosystem functioning. It is hypothesized that the relative contributions of sampled carbon sources to consumers differ between wet and dry seasons. Terrestrial producers are predicted to dominate contributions to consumers during the wet season when run-off transport these materials from the catchment and adjacent riparian zone, and reduce the availability of algal sources via scouring and sedimentation (Huryn et al., 2001). To address these hypotheses, I sampled regions with different surrounding vegetation (C3 or C4), and used $\delta^{13}C$ and $\delta^{15}N$ to identify the main sources of energy supporting consumers in different reaches of the Mara River.

2. Materials and Methods

2.1 Study area

This study was conducted on the Kenyan part of the Mara River which then crosses into Tanzania and discharges into Lake Victoria. The Mara River drains a number of forest blocks that are part of the Mau Forest Complex (MFC), which is the most extensive tropical moist broadleaf forest in East Africa (Wass, 1995). Until the middle of the past century, the 13 500 km^2 river basin was covered by montane forest in its headwaters and a mixture of shrublands and grasslands throughout its middle reaches (Mati et al., 2008, Serneels et al., 2001). Two perennial tributary rivers, the Nyangores and Amala, drain the forested headwaters and join to form the Mara mainstem (Figure 6.1). Tributaries draining the grasslands and shrublands of the middle and lower basin are ephemeral.

On the highlands, climate is relatively cool and seasonal, characterized by distinct rainfall seasons and low ambient temperatures that falls below $10^\circ C$ during January-February. Rainfall varies with altitude with the highlands receive around 1500 mm of rainfall per annum while the lowlands receive < 1000 mm. Dry conditions are experienced during December-March and August-September, while two wet seasons occur during March-May and October-December. Potential evapotranspiration varies between 1400 mm in the highlands to 1800 mm in the lowlands (Jackson and McCarter, 1994).

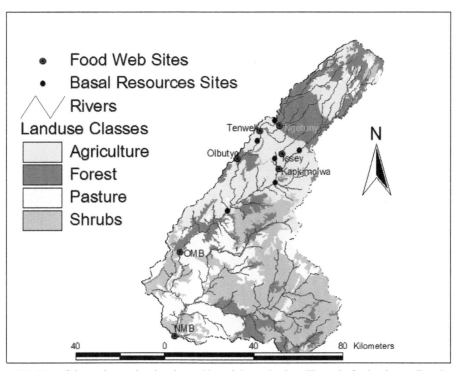

Figure 6.1. Map of the study are showing the position of the study sites. The main food web sampling sites are named while the basal resources sites are indicated with small black circles. The OMB and NMB sited were combined into Mara Main site for food web analysis.

The basin hosts substantial numbers of mammalian herbivores of varying densities. The upper reaches are under crop farming and husbandry of small herds of improved breeds of cattle. The middle reach rangelands of the Maasai contain large herds of cattle. Over 220,000 cattle are estimated to graze within the middle Mara and Talek regions (Lamprey and Reid, 2004; Ogutu et al., 2011). The savanna grasslands in Masai Mara National Reserve (MMNR) and Serengeti National Park harbour millions of resident and migratory ungulates, including 4000 hippos (*Hippopotamus amphibius*) that graze on terrestrial grasses at night and defecate into the river during the day (Lamprey and Reid, 2004; Kanga et al., 2011). Agricultural expansion over the basin has been on the increase. While cultivated land accounted for approximately 1500 km^2 of the basin in 1973, farms and tea plantations had expanded to nearly 4500 km^2 by the year 2000 (Mati et al., 2008).

2.2 Sampling protocols

Sites were selected depending on catchment land use and riparian influences by people, livestock and hippos along the river (Table 6.1). Four sites were sampled weekly for eight weeks during the dry (February-March 2013) and wet (May-July 2012) seasons to cover the temporal variation is isotopic composition of basal sources: one site (Ngetuny) was located in the forest zone where C3 vegetation

Plate 6.1: Large herbivore influences on the functioning of the Mara River, Kenya. (a) cattle drinking in the river, (b) hippos in the maistem of the Mara River, (c) accumulated cattle dung on the banks of one of the watering points in the middle reaches, and (d) water quality downstream of one of the hippo pools in one of the tributaries of the Mara River. Photos, F.O. Masese

dominates the catchment and which experiences minimal human and animal activities.; three sites (Tenwek, Issey and Kapkimolwa) more downstream were located in areas influenced by agriculture and livestock to different degrees. Three additional sites were sampled only once during the dry season (March 2013) and once during the wet season (July 2012): the Olbutyo site being influenced by both livestock grazing and agriculture, and two sites (OMB and NMB, combined and referred to hereafter as the Mara Main site) being located on the maistem Mara River within the MMNR, in river sections inhabited by large populations of hippos (Kanga et al., 2011).

Measurements of pH, dissolved oxygen (DO), temperature and electrical conductivity were done *in situ* using a YSI multi-probe water quality meter (556 MPS, Yellow Springs Instruments, Ohio, USA). Water samples were collected from the thalweg using acid washed HDP bottles for analysis of nutrients, dissolved organic carbon (DOC) and particulate organic matter (POM). For total suspended solids (TSS) and POM, stream water samples were immediately filtered through pre-weighed glass-fibre filters (Whatman GF/F, pre-combusted at 450°C, 4 h). GF/F filters holding suspended matter were carefully folded and wrapped in aluminium foil before transport in a cooler box at 4 °C to the laboratory. For water column chlorophyll *a* (Chl*a*) concentrations, a measured volume of water was filtered through a 0.7 μm pore-sized glass fibre filter, which was then wrapped in aluminium foil to

Table 6.1. Characteristics of the sampling sites used for food web studies in the Mara River basin, Kenya. Reach-scale influences capture the main land use activities in the adjoining areas of the sites while % agriculture, grasslands and forest represent the areal coverage in the catchments. River distance (RDIS) is calculated as the square root (sqrt) of drainage area (DA) upstream of the sampling site. Also presented is the density of herbivores (livestock and wildlife) within a kilometre of each sampling site.

Sampling site	Reach-scale influences	RDIS (Sqrt of DA in km^2)	% Agriculture	% Grasslands	% Forest	Herbivore density individuals/ km^2
Ngetuny	Forest	12.3	21.7	0.9	77.4	5
Issey	Agriculture & livestock	8.0	37.5	3.9	58.6	35
Tenwek	Agriculture & livestock	25.7	34.4	3.1	62.5	24
Kapkimolwa	Agriculture & livestock	26.4	41.8	16.4	41.8	60
Olbutyo	Agriculture & livestock	28.0	38.3	6.6	55.1	44
OMB	Hippos, ungulates and C4 grasslands	54.6	49.4	15.0	35.6	98
NMB	Hippos, ungulates and C4 grasslands	80.7	23.2	53.4	23.4	104

prevent exposure to light, transported on ice, and stored frozen in the laboratory prior to analysis. Both the filtered and unfiltered water samples were stored and transported in a cooler box and frozen within 10 hours of sampling.

Samples for stable isotope analyses were collected during both the wet (May-July 2011) and dry (January-April 2012) seasons. At each site samples of the dominant riparian vegetation and emergent and submerged macrophytes were collected by hand. Replicate benthic samples of coarse particulate organic matter (CPOM) were collected from pools, runs and riffles using a dip net (500 μm mesh-size). Net contents were washed with site water to remove invertebrates and inorganic materials. Samples were immediately placed in polythene bags in cooler boxes for transport to the laboratory where they were frozen until further analysis. Fine benthic organic matter (FBOM) was collected by disturbing an area by hand and filling 500 ml high density polyethene (HDPE) bottles with the mixture. Algal samples (including filamentous algae and periphyton) were collected using a scalpel. Care was taken during sample collection and processing to eliminate cross-contamination of samples. Periphyton were scrubbed from submerged surfaces (mainly slippery rocks) in riffles and runs after washing gently with site water to remove any attached invertebrates and inorganic materials. After decanting, the samples were stored in 30 ml HDPE bottles then transported to the laboratory in cooler boxes for further processing. Seston was collected by placing a 30 μm plankton net in riffles or runs at each site. After decanting and removing CPOM and other visible large fractions of material, the sample was stored in 30 ml HDPE bottles and placed in cooler boxes for transport to the laboratory.

Invertebrates were collected from riffles, runs, pools and vegetated littoral areas at each site using a dip net (500 μm mesh-size). Aquatic primary producers and macroinvertebrates were collected more frequently within eight weeks prior to fish sampling to better characterise the isotopic composition of lower trophic levels. To accommodate logistical and resource constraints, fish were collected towards the end of low- and high-flow periods. Fish were electroshocked from all sections of the river (riffles, runs, pools and around littoral habitats). Length and weight measurements were done immediately in the field for each individual fish. For SIA analysis, fish tissue was extracted from the white dorsal muscle in the field, as it is less variable in $\delta^{13}C$ and $\delta^{15}N$ than other tissue types (Pinnegar and Polunin, 1999). SIA samples were wrapped in aluminium foil and stored on ice in cooler boxes for transport to the laboratory where they were frozen until further analysis.

2.3 Sample preparation and analysis

Water quality variables determined in the laboratory included total suspended solids (TSS), particulate organic matter (POM), dissolved organic carbon (DOC), total dissolved nitrogen (TDN), total phosphorus (TP), total nitrogen (TN) and water column Chlorophyll a. GF/F filters holding suspended matter were dried (95°C) to constant weight and TSS was determined by re-weighing on an analytical balance and subtracting the filter weight. The filters were then ashed at 500°C for 4 h and re-weighed for determination of POM as the difference between TSS and ash-free-dry weight. POC was estimated from POM using a conversion factor of POM = 2*POC (Galois et al., 1996). DOC and TDN concentrations were determined using a Shimadzu TOC-V-CPN with a coupled total nitrogen analyzer unit (TNM-1). TP and TN were determined using standard colorimetric methods (APHA, 1998). Chl$a$ pigments were extracted by 90% ethanol in a water bath and concentrations were determined spectrophotometrically.

Organic samples were immediately prepared for SIA on return to the laboratory or stored in a -20°C freezer for later processing. Coarse particulate organic matter samples were thoroughly washed with de-ionized water to remove any inorganic materials and organisms. Bottom FPOM and seston were examined under a microscope to remove any living organisms. To clean periphyton the slurry was centrifuged so as to decant any heavier inorganic fractions and allow the lighter periphyton to float on the surface. After that the periphyton was decanted onto a petri dish and excess water evaporated in an oven at 60 °C for 48 h. The dry sample was then ground using a mortar and pestle, weighed and then packaged in tin cups for SIA.

Before being frozen, invertebrates were kept in polyethene bags filled with river water for at least 12 h to evacuate their guts. For confirmation, animals were then examined under a dissecting microscope to remove guts and contents. In most cases, individuals of a given taxa from each site were pooled to produce sufficient dry tissue to meet the required dry weight for SIA, except for crabs (Potamoutidae: *Potamonautes* spp.) and odonates, which were sufficiently large. To avoid contamination by carbonates, which can be enriched in ^{13}C compared with living tissues, muscle tissues were removed from crabs and mollusca, rather than using a chemical dissolution treatment which can affect $\delta^{15}N$

ratios (Carabel et al., 2006). Individual fish were analysed separately. Prior to drying, muscle samples were rinsed in distilled water and inspected for stray scales and bones. All samples were oven-dried at 60°C for 48h. Samples were then ground and weighed into tin cups for SIA.

Stable isotope analyses were performed using continuous flow EA-IRMS (elemental analyser-isotope ratio mass spectrometry). Samples were analysed on either a ThermoFinnigan DeltaPlus stable isotope mass spectrometer (Thermo Scientific) coupled to a Costech ECS 4010 EA elemental analyzer (Costech Analytical Technologies) at the Yale Earth System Center for Stable Isotopic Studies (ESCSIS, Yale University, CT, USA) or on a Thermo DeltaV Advantage coupled to a Carlo Erba EA1110 at the Katholieke Universiteit Leuven (KU Leuven, Belgium). Stable isotope ratios ($^{13}C/^{12}C$ and $^{15}N/^{14}N$) are expressed as parts per thousand (‰) deviations from standard, as defined by the equation: $\delta^{13}C$, $\delta^{15}N = [(R_{sample}/R_{reference}) - 1] \times 10^3$, where $R = ^{13}C/^{12}C$ for carbon and $^{15}N/^{14}N$ for nitrogen. The global standard for $\delta^{13}C$ is V-PDB and for $\delta^{15}N$ is atmospheric nitrogen. Each run included an internal standard interspersed within samples to provide an estimate of instrument error. For samples analyzed at ESCSIS the internal standard for animals was CBT (trout, *Salvelinus fontinalis*) muscle tissue ($\delta^{13}C = -29.0‰$, $\delta^{15}N = 15.7‰$; 12.2% N, 49.2% C) and for plants it was cocoa (*Theobroma cacao*) ($\delta^{13}C = -28.7‰$, $\delta^{15}N = 5.1‰$; 4.0% N, 47.9% C). The standard deviations of replicate samples of trout analyzed at ESCSIS were 0.29‰ for $\delta^{13}C$ and 0.09‰ for $\delta^{15}N$, and for cocoa they were 0.27‰ for $\delta^{13}C$ and 0.10‰ for $\delta^{15}N$. For KU Leuven samples, Acetanilide and Leucine were used as internal standards for $\delta^{13}C$ (two point, with blank correction) and $\delta^{15}N$ calibration. C and N contents were assessed from the TCD signal of the EA, using acetanilide (71.09% C, 10.36% N) as a standard. Replicate samples for internal standards had standard deviations of <0.3‰ for $\delta^{13}C$ and $\leq$0.2‰ for $\delta^{15}N$ during runs. Outputs from the two laboratories were compared by running 17 fish tissue samples in both. No systematic differences were obtained (paired 2-sample t-test, $t_{32} = 0.02$, $p = 0.983$ for $\delta^{13}C$ and $t_{32} = 1.05$, $p = 0.309$ for $\delta^{15}N$) and, therefore, the results from the two laboratories were combined.

For fish with molar C:N ratios >3.5, $\delta^{13}C$ were corrected for lipid content based on C:N ratios using the equation recommended by Post et al. (2007), whereby: corrected $\delta^{13}C = \delta^{13}C - 3.32 + 0.99 \times C:N$. Invertebrate $\delta^{13}C$ were not corrected for lipid content because the shifts in $\delta^{13}C$ associated with lipid removal can be very variable and taxon-specific (Logan et al., 2008).

2.4 Trophic guilds and normalization of isotope values

Classifications of fish trophic guilds were based on available literature regarding consumer diets (Corbet, 1961; Raburu and Masese, 2012) and stomach contents. *Barbus altianalis*, *B. cercops*, *B. paludinosus*, *B. kerstenii*, *B. neumayeri* and *Clarias liocephalus* were considered representative of the insectivore guild. *Labeo victorianus* and *Clarias gariepinus* represented the omnivore guild.

Chiloglanis sp. and *Bagrus dokmac* represented the piscivore guild. Macroinvertebrate trophic groups and functional feeding groups (FFGs) were based on Masese et al. (2014a) and references therein.

The isotope values of all samples were normalized to account for spatial variation in $\delta^{15}N$ and $\delta^{13}C$ that exists at the base of the food webs and prevents absolute isotope values from providing information about trophic level or the ultimate source of carbon (Cabana and Rasmussen, 1996; Vander Zanden and Rasmussen, 1999). Invertebrate herbivores (IH) have been considered as a baseline for trophic enrichment (Vander Zanden and Rasmussen, 1999; Abrantes et al., 2013). Consequently, the isotope baselines for each site and sampling occasion were obtained by averaging the mean isotope values of all invertebrate herbivores. The isotope values were then normalized by subtracting the isotope baseline of nitrogen (IBN). The advantage of using primary consumers rather than primary producers as the trophic fractionation index is that primary consumers provide reliable estimates of the isotopic ratios because they integrate both temporal and spatial variation in the isotopic signatures of primary producers (Cabana and Rasmussen, 1996; Vander Zanden and Rasmussen, 1999).

2.5 Calculation of trophic positions of consumers

This was done using the following equation (Post, 2002): $TP = \lambda + ((\delta^{15}N_{consumer} - \delta^{15}N_{baseline})/F)$, where λ is the trophic level of consumers estimating the food web base, $\delta^{15}N_{consumer}$ is the nitrogen isotopic signature of the predator being evaluated, $\delta^{15}N_{baseline}$ is the average nitrogen isotope signature of the consumers used to estimate the base of the food web, and F is the per trophic level fractionation of nitrogen, 2.54 in this case (Vanderklift and Ponsard, 2003). A $\delta^{15}N$ baseline signature was calculated independently for each study site using invertebrate herbivores, and this made it possible to compare estimates of trophic positions among sites.

Trophic enrichment/ fractionation in both ^{13}C and ^{15}N values between different consumer trophic groups and their putative food sources were determined (Bunn et al., 2013). Trophic groups and food resources considered included algae (periphyton) as a food source for all consumer groups and herbivorous invertebrates (IH) as a food source for the omnivorous and predatory consumer groups. I used herbivorous invertebrates (IH) as a food source for predatory invertebrates; the mean $\delta^{13}C$ and $\delta^{15}N$ of algae and herbivorous and predatory invertebrates (IP) as the food source (as well as the $\delta^{13}C$ and $\delta^{15}N$ of these sources individually) for omnivorous fishes; and the mean $\delta^{13}C$ and $\delta^{15}N$ of herbivorous and predatory invertebrates (as well as the $\delta^{13}C$ and $\delta^{15}N$ of these sources individually) for predatory fishes (also Bunn et al., 2013).

2.6 Data analysis

For each sampling site, river distance (RDIS) was calculated as the square root of drainage area. It has been demonstrated that the length of stream paths of tributaries leading to a point in the drainage can be expressed as a power function of the drainage area ($DA^{0.5}$, Smart, 1972). RDIS was used in this study as the independent variable against which longitudinal changes in isotopic values of basal sources were tested along the river using linear regression; relationships were estimated separately for dry and wet seasons. To capture the influence of large herbivores (mainly livestock and hippos) on basal resources in the river, the relationships between the population density of mammalian herbivores per sampling site and the stable isotopic composition of seston and FBOM was explored by simple linear regression (SLR). Herbivore (livestock, hippos and herbivorous wildlife) density was expressed as number of individuals/ km^2 in the catchment area adjoining the sampling site. Data for livestock (cattle, sheep, goats and donkeys) and wildlife (ungulates and hippos) were obtained from secondary sources that included District Development Plans for Bomet and Narok Districts (DDP, 2008a,b), Ministry of Agriculture and Livestock Production reports (MALP, 2009a,b), other unpublished reports and publications (Reid et al., 2003; Lamprey and Reid, 2004; KNBS-IHBS, 2007; KNBS-LS, 2009; Kanga et al., 2011; Ogutu et al., 2011; Kiambi et al., 2012) and were expressed as number of individuals per river kilometre. Using simple linear regression (SLR), mammalian herbivore densities were replaced by the total area of catchment under grasslands and grazing lands, which are potential sources of C4 carbon in the rivers. Seston was used as the response variable because it would help track upstream-downstream connectivity of river reaches and, thus, the influence of catchment land use and livestock. FBOM was chosen to cater for the settling out of excreta and terrestrial OM into the benthos and would, thus, help assess local influences.

Stable Isotope Analysis in R (SIAR), a Bayesian mixing model that runs in R (R Project for Statistical Computing, Vienna, Austria) was used to estimate the contributions of different basal sources to consumer diets (Parnell et al., 2010). As opposed to other commonly used mixing models (e.g., IsoSource; Phillips and Greg 2003), SIAR accounts for variability and uncertainties associated with natural systems to give a more reliable estimate of the dietary composition of consumers (Parnell et al., 2010). The SIAR model considers available sources to produce a range of feasible solutions while taking into account uncertainty and variation both in consumer and trophic enrichment factors (TEF). The model also provides error terms (the residual error) that give information on the variability that cannot be explained based on diet alone (Parnell et al., 2010). Models were run for each site and season separately for the different trophic guilds (for invertebrates) and/ or individual taxa (for fish). Prior to analyses, consumer and source data were plotted to ensure that consumers were within the isotopic mixing space.

Only C3 and C4 producers and periphyton were included in the models as possible sources of energy. Lichens occurred at Ngetuny, Issey and Kapkimolwa, but their isotopic compositions overlapped with that of periphyton at the sites and were therefore combined. Macrophytes were not included because they occurred in very low and patchy densities at the sampled river reaches and, hence, assumed to contribute very little to food webs. For periphyton, mean $\delta^{13}C$ and $\delta^{15}N$ of sources, and respective standard deviations, were calculated for each site and season to eliminate possible sources of error arising from inter-site variability in isotopic composition (Finlay, 2004). For C3 and C4 sources, data were pooled from all sites, as terrestrial organic matter (OM) from the overall catchment is transported into the system.

Because different trophic groups were used in the models, TEFs were set to zero and, instead, the consumers' stable isotope composition was corrected for trophic fractionation before the analyses. Trophic shifts used for ^{15}N (in relation to the first trophic level) were 0.6 ± 1.7‰ for herbivorous invertebrates, 1.8 ± 1.7‰ for predatory invertebrates, 4.3 ± 1.5‰ (±SD) for omnivorous fish and 5.7 ± 1.6‰ for predatory fish (Bunn et al., 2013). To account for trophic shifts in δ^{13}C, 0.5‰ was used (McCutchan et al., 2003) but a large TEF SD of 1.3‰ were set to account for uncertainty in these fractionation values (e.g., Post, 2002). Concentration dependencies were set to zero. Where values were single measurements, the SIAR model for source contributions to the consumer were run on the SIARSOLO code. SLR models were used to explore relationships between the proportion of C4 vegetation cover in the catchments (expressed in %) and the importance of C4 and C3 producers to consumer groups in the river.

3. Results

3.1 Physical and chemical variables

There were both seasonal and spatial variability in physical and chemical variables (Plate 6.2, Table 6.2). At all sites, except the Mara Main site, DOC concentrations were higher during the wet than during the dry season. Similarly, at most sites TDN, TSS and TP were higher during the wet season compared with the dry season, while TN, conductivity and temperature were lower during the wet season (Table 6.2). At Tenwek, Issey and Kapkimolwa sites, % POM reduced during the wet season indicating input of sediments poor in organic matter. Chlorophyll a was lowest at the forest Ngetuny (3.2±1.1 μg/L) and highest at Olbutyo site (34.3±9.2 μg/L). Dry season mg POC: mg Chlorophyll a values ranged from 86±33 at Olbutyo to 1418±206 at the Ngetuny site (Table 6.2). Overall, the agriculture, livestock and hippo influenced sites were warmer, with higher specific conductivity, suspended sediments and concentrations of nutrients and Chlorophyll a than the forest Ngetuny site.

3.2 Basal resources

Wide scatter in δ^{13}C and δ^{15}N values for basal resources (C3 and C4 terrestrial plants, CPOM, FPOM, seston, biofilm, lichens, periphyton (diatoms), filamentous algae and cyanobacteria) were observed during the dry (Figure 6.2) and wet (Figure 6.3) seasons. The mean δ^{13}C values of the main producer categories collected at different sites were well differentiated (Table 6.3). However, the δ^{13}C values of C3 and C4 producers did not differ among sites and seasons (data not shown), so values from all sites and seasons were combined and the average used in the SIAR models. δ^{13}C of periphyton was lower

Plate 6.2: Seasonal variability in flows in the Mara River; (a) Tenwek site during the wet season and (b, c, d) Old Mara Bridge site during (b) wet season, (c, d) dry season. Increased primary production in backwaters was noted (d) during the dry season the Old Mar Bridge site.

during the wet season compared with the dry season at all sites. Longitudinally, $\delta^{13}C$ values were lowest at the Ngetuny forest site (-25.8±1.8‰ in the dry and -26.9±1.1‰ in the wet season) and highest at the Mara Main site (-17.3±0.9‰ in the dry and -20.0±0.7‰ in the wet season). The $\delta^{15}N$ values of C4 producers were generally higher than those of C3 producers (Table 6.3). For periphyton, $\delta^{15}N$ values ranged from 6.3‰ at Olbutyo to >9‰ at Issey. There were longitudinal increases in $\delta^{13}C$ of seston, periphyton, CPOM, FBOM and lichens in the Mara River during both the dry and wet seasons, but not for filamentous algae (Figure 6.4). For sites in agricultural areas with animal inputs of C4 material (Issey and Kapkmolwa), $\delta^{13}C$ was higher during the dry season. During both the dry and wet seasons, CPOM had higher $\delta^{13}C$ values at the OMB (mean ± SD, -17.5±0.8 ‰ during the dry season, -15.5±0.8 ‰ during the wet season) and NMB (-16.3±0.5 ‰ during the dry season; -18.5±1.0 ‰ during the wet season) sites that received subsidies of C4 grasses by hippos (Figure 6.4).

Longitudinal changes in $\delta^{15}N$ in the river did not follow those of $\delta^{13}C$ for most of the basal resources (Figure 6.5). FBOM, periphyton, filamentous algae and lichens did not display significant longitudinal trends, but for FBOM the relationship was only marginally significant (R^2 =0.22, p = 0.09). However, $\delta^{15}N$ values of these sources were higher at agriculture and livestock influenced sites (Figure 6.5). Seston and CPOM displayed significant longitudinal trends during the dry season (seston also during

Table 6.2. Mean (±SD) physical and chemical variables at each site and season. DOC = dissolved organic carbon, POM = particulate organic matter, TSS = total suspended sediments, TDN = total dissolved nitrogen, DO = dissolved oxygen, TP= total phosphorus, TN = total nitrogen.*

Sites	Land use influences	Season	n	DOC (mg/L)	% POM[#] in TSS	TDN (mg/L)	pH	DO (%)	Specific conductivity (μS/cm)
Ngetuny	Forest	Dry	12	2.3±1.2	21.2	0.5±0.3	6.7±0.3	73.7±3.3	51.3±3.9
		Wet	12	1.9±0.4	17.5	0.3±0.6	6.7±0.2	78.4±3.7	57.3±5.5
Issey	Agriculture & livestock	Dry	12	4.3±0.7	13.7	1.1±1.0	7.6±0.7	76.8±14.6	151.7±37.6
		Wet	12	6.2±3.2	9.5	0.9±0.4	7.1±0.2	72.8±5.1	133.7±36.5
Kaplimolwa	Agriculture & livestock	Dry	12	2.8±0.9	7.8	1.3±0.8	7.5±0.3	76.3±17.3	163.8±16.5
		Wet	12	4.3±2.1	6.3	1.1±0.7	7.3±0.5	87.0±11.4	107.6±53.8
Tenwek	Agriculture	Dry	12	2.4±0.6	8.3	0.7±0.2	7.5±0.6	72.1±4.7	63.8±8.2
		Wet	12	2.8±1.4	7.6	0.9±0.3	6.9±0.2	84.3±13.0	58.3±11.6
Olbutyo	Livestock & agriculture	Dry	3	2.9±0.7	8.9	1.6±0.3	7.2±0.4	78.7±21.3	70.2±11.2
		Wet	3	3.5±3.2	12.3	2.1±0.2	7.4±0.7	82.6±11.3	60.1±23.5
Mara Main	Hippos, ungulates and savanna	Dry	6	6.3±4.5	11.2	1.0±0.8	6.3±1.0	67.2±7.3	374.2±117.8
		Wet	6	3.8±6.0	16.9	1.1±0.7	7.7±0.9	80.4±10.7	106.0±132.7

Table 6.2. continued

Sites	Land use influences	Season	n	Temperature (°C)	TSS (mg/L)	TP (mg/L)	TN (mg/L)	mgPOC:mgChla[⁑] (μg/L)
Ngetuny	Forest	Dry	12	14.2±1.3	42.8±8.2	0.25±0.4	0.9±0.6	1418±206
		Wet	12	14.4±2.4	61.3±11.2	0.36±0.1	1.1±0.8	-
Issey	Agriculture & livestock	Dry	12	19.8±2.7	57.1±11.5	0.15±0.1	2.8±0.8	253±45.2
		Wet	12	19.5±1.5	83.0±12.4	0.38±0.3	1.2±0.4	-
Kaplimolwa	Agriculture & livestock	Dry	12	19.9±2.7	64.6±18.4	0.12±0.1	1.6±0.7	1073±24
		Wet	12	20.3±3.5	114.5±37.5	0.24±0.2	1.3±0.6	-
Tenwek	Agriculture	Dry	12	19.5±1.6	57.3±14.2	0.24±0.3	1.9±1.2	449 ±102
		Wet	12	18.6±1.8	104.8±23.4	0.19±0.1	1.0±0.4	-
Olbutyo	Livestock & agriculture	Dry	3	19.2±2.0	65.3±7.7	0.28±0.1	2.0±0.5	86±33
		Wet	3	16.9±2.0	93.3±12.6	0.39±0.1	1.3±0.8	-
Mara Main	Hippos, ungulates and savanna	Dry	6	23.9±2.6	72.9±16.2	0.29±0.2	1.5±1.0	183±26
		Wet	6	18.4±3.3	126.7±43.7	0.60±0.3	1.7±1.1	-

*Notes. POC is derived from POM using the following conversion: POM=2*POC (Galois et al., 1996); [⁑]water column chlorophyll a (Chla) was determined during the dry season only, $n = 3$.

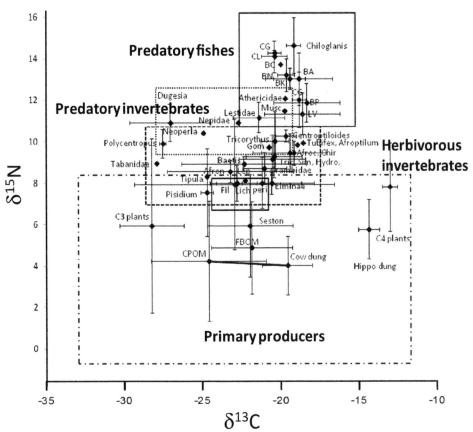

Figure 6.2. Mean ± 1 SD) for $\delta^{13}C$ and $\delta^{15}N$ showing the spread and fractions of the various basal resources and consumers in the Mara River and its tributaries during the dry season. Primary producers, invertebrate herbivores, invertebrate predators, and insectivorous fishes trophic groups are enclosed in a box but show large overlap. CPOM = coarse and particulate organic matter, FPOM = fine particulate organic matter, primary producers- fil = filamentous algae, lich = lichens, Peri = periphyton; invertebrates- Lep = *Lepidostoma*, Afron = *Afronurus*, Tric = *Tricorythus*, Sim = Simuliidae, Hydro = Hydropsyche, Afroc = *Afrocaenis*, Chir = Chironominae, Gom = Gomphidae, Musc = Muscidae; fishes- LV = *Labeo victorianus*, BC = *Barbus cercops*, BP = *B. paludinosus*, BA = *B. altianalis*, BK = *B. kerstenii*, CG = *Clarias gariepinus*, CL = *C. liocephalus*.

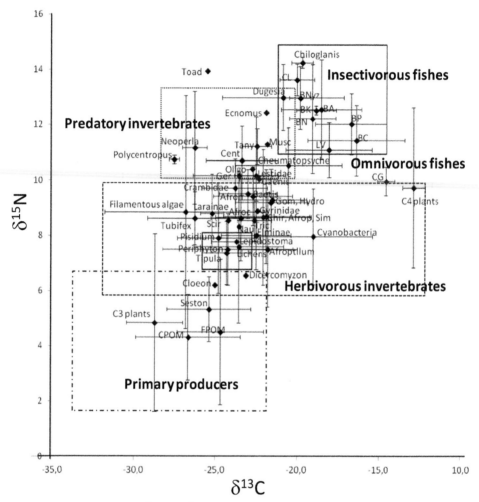

Figure 6.3. Means (± 1 SD) for δ¹³C and δ¹⁵N showing the spread and fractions of the various basal resources and consumers in the Mara River during the wet season. Trophic levels that include primary producers, invertebrate herbivores, invertebrate predators, omnivorous fishes and insectivorous fishes are enclosed in a box but show large overlap. CPOM = coarse and particulate organic matter, FPOM = fine particulate organic matter; invertebrates- Scir = Scirtidae, Nau = Naucoridae, Afron = *Afronurus*, Tric = *Tricorythus*, Sim = Simuliidae, Hydro = Hydropsyche, Afroc = *Afrocaenis*, Chir = Chironominae, Gom = Gomphidae, Oligo = Oligoneuridae, Cent = Centroptiloides, Tany = Tanypodinae, Musc = Muscidae; fishes- LV = *Labeo victorianus*, BC = *Barbus cercops*, BP = *B. paludinosus*, BA = *B. altianalis*, BK = *B. kerstenii*, BNyz = *B. nyanzae*, CG = *Clarias gariepinus*, CL = *C. liocephalus*.

138

Table 6.3. $\delta^{13}C$ and $\delta^{15}N$ (mean ± SD; in ‰) of the main producer categories collected at different sites and in the surrounding catchment in the dry and wet seasons. n is the number of samples used to calculate the mean.*

Site	Land use influences	Producers	Dry			Wet		
			n	$\delta^{13}C$	$\delta^{15}N$	n	$\delta^{13}C$	$\delta^{15}N$
All sites		C3¥	61	-28.5±1.5	4.1±2.8	61	-28.5±1.5	4.1±2.8
		C4¥	8	-12.9±0.6	8.4±2.9	8	-12.9±0.6	8.4±2.9
Ngetuny	Forest	C3	1	-28.0	3.0		NC	NC
		C4		NA	NA		NA	NA
		Periphyton	11	-25.8±1.8	6.8±0.5	13	-26.9±1.1	6.5±1.1
Issey	Agriculture	C3		NC	NC	21	-28.4±1.7	5.5±3.7
	& livestock	C4		NC	NC	2	-13.4±0.5	13.1±1.4
		Periphyton	8	-22.9±1.0	9.1±0.8	18	-23.1±2.0	9.2±1.6
Kapkimolwa	Agriculture	C3		NC	NC	12	-28.3±1.3	4.3±3.5
	& livestock	C4		NA	NA		NA	NA
		Periphyton + Lichens	15	-19.6±2.6	8.4±0.5	18	-22.6±1.7	7.5±0.8
Tenwek	Agriculture	C3	9	-28.4±1.3	6.9±5.7		NC	NC
		C4	1	-12.7	8.3		NC	NC
		Periphyton	14	-20.2±2.3	8.0±0.7	15	-22.3±2.3	7.3±0.6
		Lichens	3	-23.4±0.7	6.6±0.4		NC	NC
Olbutyo	Livestock	C3	5	-29.7±1.3	5.3±3.7	5	-28.0±0.6	4.2±2.0
	&	C4	1	-13.0	8.7		NC	NC
	agriculture	Periphyton	2	-20.6±0.2	6.3±0.1	2	-26.1±1.3	6.3±0.4
Mara Main	Hippos &	C3	8	-27.9±2.4	4.8±2.3		NC	NC
	savanna	C4	4	-13.3±1.3	6.1±1.1		NC	NC
	grasslands	Periphyton	6	-17.3±0.9	6.4±1.2	2	-20.0±0.7	6.4±0.1

*Notes: NA, not available; NC, not collected; ¥Samples from different sites and seasons combined.

the wet season) with $\delta^{15}N$ decreasing with river distance from source (Figure 6.5). Sites under livestock and agriculture influences had higher seston and CPOM $\delta^{15}N$ values, with low values in the forested upper reaches and in the savanna grasslands in the MMNR where hippos had an influence. The % of agricultural land use was a strong predictor (R^2 =0.65, $p < 0.05$) of $\delta^{15}N$ values of FBOM in the study area during the dry season.

Both herbivore density and the estimated % of C4 vegetation cover in catchments had significant relationships with $\delta^{13}C$ values of seston and FBOM along the river (Figure 6.6). The relationships were significant during both the dry and wet seasons, and $\delta^{13}C$ values were higher at livestock-influenced agriculture sites during the dry season (Figure 6.6). The % of forest land use displayed a significant negative relationship with the $\delta^{13}C$ values of seston during the dry (R^2 =0.85, $p < 0.01$) and wet (R^2 =0.67, $p < 0.05$) seasons.

3.3 Consumer groups

3.3.1 Normalizations of isotope values

Variability in basal resource signatures was evident across sites and sampling occasions for the IH used for normalizing base $\delta^{15}N$ values. Because of widespread omnivory, as depicted by large error bars, some predators (e.g. Libellulidae, Gomphidae) were classified among herbivores and vice versa

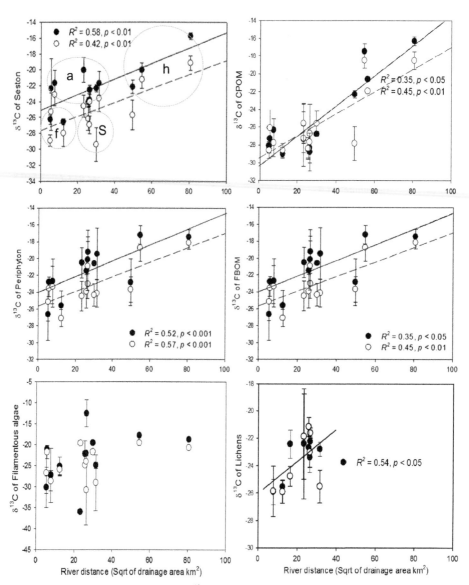

Figure 6.4: Spatio-temporal variability in the δ¹³C of seston, CPOM, periphyton, FBOM, filamentous algae and lichens in the Mara River and its tributaries. Lines are linear regression relationships; dotted lines and open circles = wet season, full line and shaded circles = dry season. Single measurements per site do not have error bars. Dotted circles enclose sites under similar influences and apply to all panels- a = agriculture and livestock, f = forest, s = savanna livestock and agriculture and h = hippos and savanna grasslands.

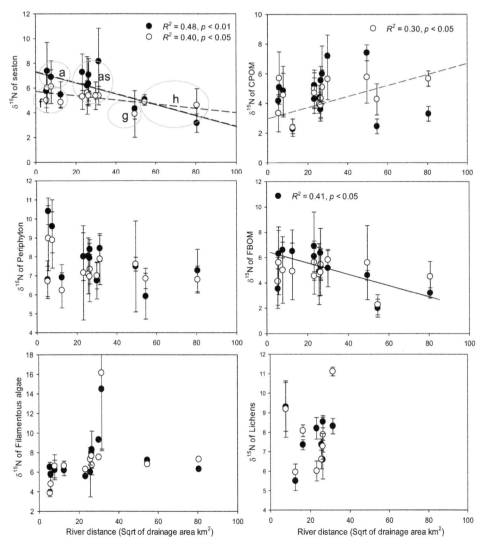

Figure 6.5: Spatio-temporal variability in the δ¹⁵N of seston, CPOM, FBOM, periphyton, filamentous algae and lichens in the Mara River and its tributaries. Lines are linear regression relationships; dotted lines and open circles = wet season, full line and shaded circles = dry season. Single measurements per site do not have error bars. Dotted circles enclose sites under similar influences and apply to all panels- a = agriculture and livestock, f = forest, s = savanna livestock and agriculture, g =savanna grazing rangelands and h = savanna grasslands and hippos.

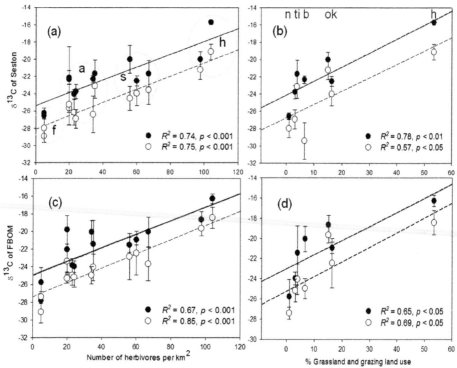

Figure 6.6: Relationships between number of herbivores per km^2 and δ^{13}C of (a) seston and (c) fine benthic organic matter (FBOM), and relationships between the percentage of grassland and pasture land use and δ^{13}C of (b) seston and (d) FBOM in the Mara River and its tributaries. Lines are significant linear regression relationships. Full lines and shaded circles are for dry season, dotted lines and open circles are for the wet season. Dotted circles enclose sites under similar influences and apply to panels (a) and (c)- a = agriculture and livestock, f = forest, s = savanna livestock and agriculture and h = savanna grasslands and hippos; small letter across panel (b) are for site names in Table 1 and also apply to panel (d): n = Ngetuny, t = Tenwek, I = Issey, b = Olbutyo, o = OMB, k = Kapkimolwa, and h = NMB.

(Figure 6.7). Further analysis of macroinvertebrate and fish trophic groups and their putative food resources showed different trophic enrichment/ fractionation in both ^{13}C and ^{15}N values (Table 6.4). For macroinvertebrates, there were overlaps between groups separated by gut content analysis (Masese et al., 2014a).

3.3.2 Trophic positions

The trophic positions (TrPos) for fish ranged between 4 and 6 without a clear distinction among species and sites (Figure 6.8). However, the limited distribution of fish species across sites limited spatial comparisons. *Clarias liocephalus* (insectivore) was the most widespread species by occurring at five of the six sites and recorded the highest TrPos at the sites it was captured. *C. warneri* and *Bagrus docmak* were the least widely distributed, occurring at one site each; both had TrPos of >5. TrPos for macroinvertebrate separated clearly between predators (values between 3.5 and 4.5) and herbivores (values between 1.5 and 3). TrPos for predatory invertebrates were within ranges for

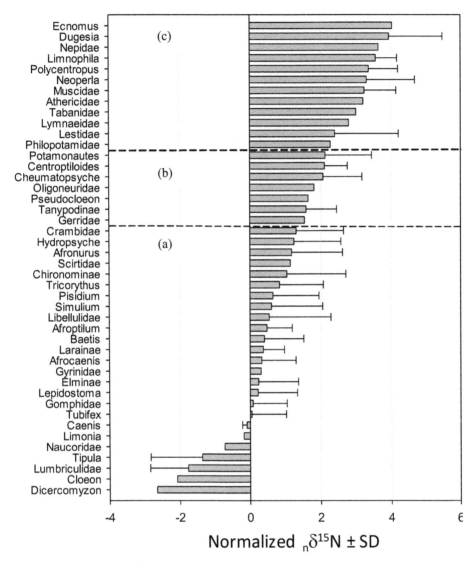

Figure 6.7. Normalized $\delta^{15}N$ ($_n\delta^{15}N$) values (mean ± SD) of the invertebrate taxa from the Mara River, Kenya, and its tributaries. Normalization was done by subtracting the absolute value of individual taxa with the average value of all invertebrate herbivores (IH) at each sampling site during each sampling occasion. The horizontal lines separate trophic groups according to Masese et al., 2014a; a) mainly herbivores, b) mainly omnivores, and c) mainly predators. Taxa without error bars are single measurements.

Table 6.4: Means (±SD) and 95% credibility intervals of diet–tissue $\delta^{13}C$ and $\delta^{15}N$ fractionation for consumers from the Mara River, Kenya, and its tributaries. Fractionation (Δ) values have been calculated as the mean difference in $\delta^{13}C$ and $\delta^{15}N$ between consumer groups and their putative food sources.

Transfer of food source to consumer	No. of trophic levels	n	$\Delta^{13}C$ Mean±SD	$\Delta^{13}C$ 95% CI	$\Delta^{15}N$ Mean±SD	$\Delta^{15}N$ 95% CI
A to IH	1	256	1.1±1.0	-0.8-2.4	1.4±0.7	0.1-2.3
A to IP	2	59	1.5±2.5	-3.1-5.8	2.7±1.7	-1.1-5.7
IH to IP	1	58	0.9±1.7	-1.4-3.2	1.6±1.7	-1.4-4.5
A to OF	1	110	2.7±2.2	-0.6-6.7	3.8±1.1	2.0-5.5
A to IF	2 to 3	372	2.7±2.1	0.1-5.9	5.4±1.2	3.5-7.1
A to PF	3 to 4	7	3.7±1.3	0.9-5.0	6.5±0.7	5.1-7.1
IH to OF	1	110	0.4±2.2	-2.9-4.0	2.5±1.1	0.8-4.6
IH to FP	2 to 3	7	2.3±1.7	-0.2-4.2	5.7±0.6	5.1-6.9
IH to IF	1 to 2	372	1.2±2.3	-2.0-4.1	4.2±1.3	1.7-6.1
IP to OF	1	110	1.0±2.0	-2.7-4.3	2.0±1.6	0.2-6.2
IP to IF	1	372	1.6±1.6	-0.7-4.1	2.8±0.9	1.3-4.3
Mean (IH,IP) to IF	1	372	1.0±1.8	-1.8-4.0	4.0±0.9	2.7-5.2
Mean (A,IH,IP) to OF	1	110	1.7±2.0	-1.1-5.2	3.1±1.2	1.0-5.3

A, algae (mostly periphyton); IH, herbivorous invertebrates; IP, predatory invertebrates; OF, omnivorous fishes; IF, insectivorous fishes; PF, piscivorous fishes.

insectivorous fishes (e.g., *B. altinialis*, *C. liocephalus*, *B. neumayeri*). At the agriculture and livestock influenced sites (Tenwek, Issey, and Kapkimolwa) mean TrPos were slightly higher for all functional feeding groups compared with the forest Ngetuny and hippo influenced Mara Main sites (Figure 6.8).

3.3.3 Trophic guilds

Different consumer trophic guilds had different $\delta^{13}C$ and $\delta^{15}N$ values across sites. Dry season values were generally higher than wet season values (Table 6.5). Ngetuny site recorded the lowest $\delta^{13}C$ values (range, -26.9 to -24.7) for macroinvertebrate guilds, and these were within values recorded for periphyton and C3 producers. The $\delta^{13}C$ value of -22.8 recorded for insectivorous fish in the dry season was lowest in the study area. Similarly, Tenwek site had low $\delta^{13}C$ values for filterers (-25.6), scrapers (-22.6±1.9) and shredders (-25.0) during the wet season (Table 6.5). At Issey and Kapkimolwa sites, most dry season $\delta^{13}C$ values for macroinvertebrate trophic guilds were higher than wet season values, while there were no changes in $\delta^{15}N$ values between the seasons (Table 6.5). However, insectivorous and omnivorous fishes did not display any seasonal variability in their $\delta^{13}C$ and $\delta^{15}N$ values. At the Mara Main site, consumers had the highest $\delta^{13}C$ values (range -20.3 to -14.5‰) and in general $\delta^{13}C$ values were higher in the wet season (Table 6.5), while $\delta^{15}N$ values were lower than at rest of the sites, only comparable to those recorded at the forest Ngetuny site.

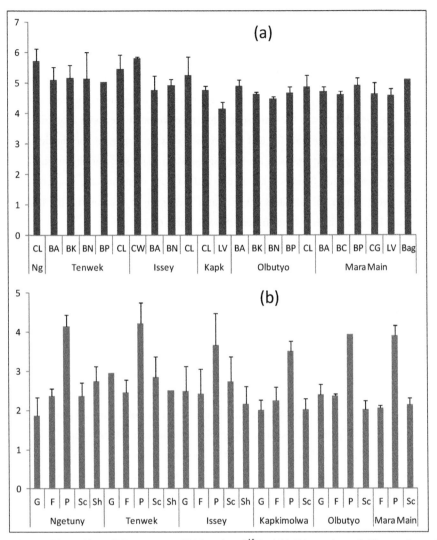

Figure 6.8. Trophic positions (TrPos, mean ± SD) based on $\delta^{15}N$ of (a) fish species and (b) macroinvertebrate functional feeding groups from different sites in the Mara River basin, Kenya. Ng = Ngetuny, Kapk = Kapkimolwa, CL = *Clarias liocephalus*, BA = *Barbus altianalis*, BK = *B. Kerstenii*, BN = *B. neumayeri*, BP = *B. paludinosus*, CW = *C. warneri*, LV = *Labeo victorianus*, BC = *B. cercops*, CG = *C. gariepinus*, Bag = *Bagrus docmak*, G = gatherers, F = filterers, P = predators, Sc = scrapers, Sh = shredders. Taxa and groups without error bars are single measurements.

3.4 Stable isotope mixing models

Bayesian mixing model results are graphically presented for macroinvertebrate scrapers and insectivorous fishes for illustration (Figure 6.9). Overall, model results indicate that periphyton dominated contributions to macroinvertebrates and fishes in the Mara River (Figure 6.10 for illustration). The importance of terrestrial C3, C4 and autochthonous production (mainly periphyton) differed between consumer groups, sites and seasons (Table 6.6). During the dry season, periphyton and C3 producers were the main sources for invertebrate filterers at the forest site with 25-86% (95% credibility interval (CI)) contribution, while the three sources (C3, C4, and periphyton) were all important for insectivorous fishes (Table 6.6). At Tenwek site, periphyton and lichens were the main sources for filterers (5-51% and 8-53%, respectively) and insectivorous fishes (20-57% and 23-62%, respectively). At Issey and Kapkimolwa sites periphyton was the main source for most consumers. At the Olbutyo site, periphyton was the main source for insectivorous fishes with a contribution of 28-75%. All sources (periphyton and C3 and C4 producers) were important for shredders, filterers, gatherers and predatory invertebrates at the site. At the Mara Main site, periphyton and C4 sources were equally important for filterers with 3-76% and 11-75% contributions, respectively. Source contributions were similar for predatory invertebrates during the dry season but the importance of periphyton was higher for insectivorous fishes (28-75%, Table 6.6).

During the wet season, notable shifts were observed with C4 producers dominating sites influenced by higher numbers of livestock and hippos (Table 6.6). For most consumer groups at the forest Ngetuny site, C3 producers and periphyton were important, although the 95% CI of some contributions included 0%. At Issey site, the importance of periphyton was lower in the wet than in the dry season for filterers and insectivorous fishes while that of C4 producers was higher; the importance of other sources for the rest of the consumer groups displayed mixed patterns. However, for most of the consumer groups, C4 producers were important during the dry season while C3 producers were important during the wet season (Table 6.6). During the wet season C3 producers were the main source for shredders (21-63%). At Kapkimolwa site, the importance of C3 and C4 producers increased for all consumers during the wet season, except for insectivorous fishes where C4 contributions reduced from 17-37% to 5-38% (Table 5). At Olbutyo site, periphyton (27-54%) and C4 producers (44-52%) were equally important for insectivorous fishes. At the Mara Main site, C4 sources dominated for all consumers during the wet season (Table 6.6). The importance of C4 producers for insectivorous fishes increased from 1-37% to 57-82% from the dry season to the wet season, while that of periphyton reduced from 40-94% to 1-41%. Similarly the importance of C4 producers for omnivorous fishes increased from 10-42% to 51-77% while that of periphyton reduced from 40-84% to 9-48%.

There were significant positive relationships ($p < 0.05$) between the estimated proportion of C4 vegetation cover in the catchments and the importance of C4 producers to macroinvertebrate collector-filterers, collector-gatherers and scrapers during the dry and wet seasons and predatory macroinvertebrates and insectivorous fishes during the wet season as estimated by the Bayesian mixing models (Figure 6.11). The SLR relationships were negative with the estimated importance of C3 sources to macroinvertebrate collector-filterers, scrapers, predators and insectivorous fishes during the wet season and with scrapers during the dry season (Figure 6.11). For macroinvertebrate collector-gatherers, there were no significant relationships with the estimated importance of C3 sources.

Table 6.5. Mean consumer (±SD) δ^{13}C and δ^{15}N for macroinvertebrates and fish trophic guilds at each site and season. For invertebrates, numbers in parentheses after the δ^{13}C values are the number of samples/ replicates, while for fish they are the number of species followed by the number of individuals per sample. NA, not available; NC, not collected.

Guild	Season	Issey	Kapkimolwa	Mara Main	Ngetuny	Olbutyo	Tenwek
Invertebrates δ^{13}C							
Gatherers	Dry	-20.0±0.5 (7)	-18.7±1.5 (5)	NC	-25.8 (1)	-20.7±2.5 (4)	NC
	Wet	-22.7±2.6 (9)	-19.7±1.9 (3)	NC	-26.9 to -25.8 (2)	NC	-25.6 to -21.4 (2)
Filterers	Dry	-21.4±1.7 (20)	-20.5±0.8 (14)	-15.3 (1)	-24.8±1.5 (6)	-19.8±0.7 (3)	-19.7±0.6 (9)
	Wet	-22.1±1.7 (8)	-20.3±0.7 (3)	-14.6 (1)	NC	NC	-20.6±2.3 (3)
Predators	Dry	-23.2 to -23.0 (2)	-20.3±0.7 (3)	-20.3 to -17.2 (2)	NC	-20.8 (1)	NC
	Wet	-21.5±1.7 (11)		-14.7 to -13.8 (2)	-26.2±2.1 (3)	NC	-22.4 to -22.1 (2)
Scrapers	Dry	-23.3 to -21.4 (2)	-20.3±1.9 (5)	-17.0 to -15.7 (2)	NC	-21.5±1.8 (4)	NC
	Wet	-24.8 (1)	-21.0 (1)	-16.8 to -15.9 (2)	-25.4±1.2 (4)	NC	-22.6±1.9 (5)
Shredders	Dry	-22.7 (1)	NC	NC	NC	NC	NC
	Wet	-22.8±1.1 (4)	NC	NC	-24.7±0.4 (7)	NC	-25.0 (1)
Fish						-19.3±0.8 (4, 24)	
Insectivores	Dry	-20.6±0.7 (4, 34)	-19.7±0.8 (2, 18)	-16.1±2.5 (2, 23)	-22.8 (1, 1)	-18.9±1.1 (5, 29)	-20.5±1.0 (4, 32)
	Wet	-19.6±0.8 (2, 14)	-19.7±0.8 (2, 12)	-14.5±1.5 (3, 27)	-18.9 (1, 1)		-19.8±0.9 (4, 24)
Omnivores	Dry	NA	-19.9±2.2 (1, 14)	-16.6±1.1 (2, 25)	NA	NA	NA
	Wet	NA	-20.0±0.7 (1, 9)	-14.9±1.2 (2, 21)	NA	NA	NA
Piscivores	Dry	NA	NA	-14.8 (1, 1)	NA	NA	NA
δ^{15}N							
Invertebrates							
Gatherers	Dry	9.9±0.4	9.2±1.3	NC	6.7	9.1±0.7	NC
	Wet	9.6±1.7	9.3±1.0	NC	5.0 to 6.7	NC	9.0 to 10.2
Filterers	Dry	9.9±0.9	9.4±0.9	6.5	7.1±0.5	9.0±0.2	9.0±0.5
	Wet	9.2±1.0	9.2±0.8	6.3	NC	NC	10.2±2.1
Predators	Dry	10.8 to 11.6	10.5±1.3	8.9 to 9.3	NC	10.5	NC
	Wet	10.3±1.6	NC	7.7 to 8.5	9.1±0.7	NC	10.4 to 10.6
Scrapers	Dry	7.2 to 10.9	8.8±0.7	6.5	NC	8.1±0.6	NC
	Wet	9.5	9.4	6.3 to 7.3	7.1±0.9	NC	8.7±1.3
Shredders	Dry	7.9	NC	NC	NC	NC	NC
	Wet	7.6±1.3	NC	NC	8.3±0.8	NC	7.8
Fish							
Insectivores	Dry	14.2±0.7	13.2±0.8	11.1±0.6	12.4	13.2±0.3	13.4±1.0
	Wet	13.3±0.7	13.1±1.3	10.7±0.5	13.8	13.±0.3	13.2±0.4
Omnivores	Dry	NA	11.8±0.7	10.6±0.7	NA	NA	NA
	Wet	NA	11.6±0.3	10.0±0.5	NA	NA	NA
Piscivores	Dry	NA	NA	11.6	NA	NA	NA

4. Discussion

Studies that address energy flow in riverine food webs are important to identify specific habitats and energy sources that underpin productivity (Naiman et al., 2012). The SIAR model results in this study show the relative importance of different sources of energy for consumers in the Mara River which are spatially and seasonally variable. Overall, periphyton was found to be the major source of energy at most of the sites and for consumers during the dry season while the importance of C3 relative to C4 producers reduced at agricultural and hippo influenced sites (Figure 6.10). In contrast, a range of other studies have suggested that C4 producers contribute minimally as an energy and nutrient source to consumer biomass as compared with C3 and autochthonous (algae and periphyton) producers (Clapcott and Bunn, 2003; Roach, 2013). Moreover, most studies indicate that algal carbon, including periphyton, is the predominant production source (Delong and Thorp 2006; Roach, 2013).

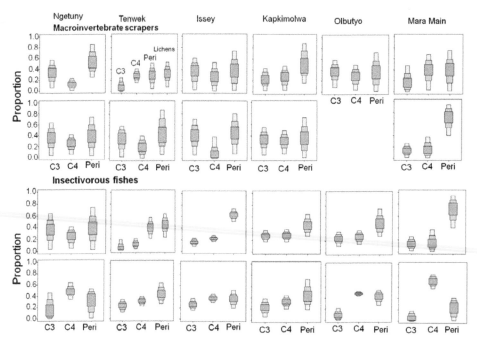

Figure 6.9. Bayesian SIAR mixing model outputs for the proportions of C3, C4, lichens and periphyton (peri) producers for macroinvertebrate-scrapers (except at Tenwek where macroinvertebrate-filterers are presented) and insectivorous fishes from the different sites (Ngetuny-Mara Main) during the dry (upper panels of each consumer group) and wet (lower panels for each consumer group) seasons. Boxes indicate the 50%, 75% and 95% Bayesian credibility intervals. At Olbutyo, scrapers were collected during the dry season only.

The importance of C4 terrestrial producers relative to autochthonous and C3 producers is attributable to importation of terrestrial production by herbivores into the Mara streams and rivers. High $\delta^{13}C$ values of OM at agricultural sites (Figure 6.4) indicate C4 carbon sources arising from livestock in agricultural catchments (Plate 6.1). Grassland vegetation is limited in the forest and along riparian areas (Table 6.1), implying that large herbivores transport organic matter to streams and rivers during the dry season (Plate 6.1). The $\delta^{15}N$ values for basal resources were also higher at the agriculture and livestock influenced sites (>7‰ and as high as 13‰) compared with the forest and hippo influenced sites (<7‰, Figure 6.5), suggesting that a significant fraction of the nitrate is from agriculture and animal activities (Harrington et al., 1998; Anderson and Cabana, 2005). In the protected areas of the MMNR, large populations of hippos that number over 4000 graze on terrestrial savanna C4 grasses during the night and then excrete partially digested grasses in the river during the day (Kanga et al., 2011). With a single hippo able to transfer approximately 9t of wet weight terrestrial organic matter into aquatic ecosystems yearly (Heeg and Breen 1982; Naiman et al., 2005), the large number of hippos in the Mara River represent approximately 800 kg of wet weight of terrestrial subsidy input per river kilometre per day. In contrast, a Madagascan catchment lacking large herbivores, low amounts of C4 carbon enter the river systems than expected based on areal cover (Marwick et al., 2014b).

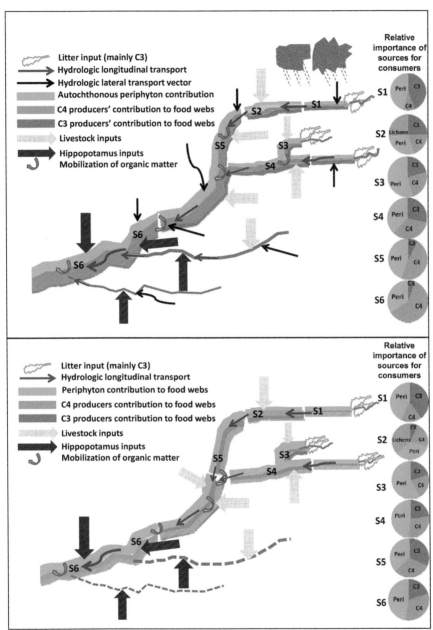

Figure 6.10. SIAR estimated dry (bottom) and wet (top) season contributions of C3 and C4 producers, lichens and periphyton (peri) to consumers in the Mara River. The width of the colour is proportional to the contributions of the four sources of carbon to macroinvertebrates and fishes in the immediate river reaches. The pie charts represent the proportions averaged for different consumer groups per site (S1 = Ngetuny, S2 = Tenwek, S3 = Issey, S4 = Kapkimolwa, S5 = Olbutyo, S6 = Mara Main. The grey lines represent the river network not considered during this study; the dotted lines represent seasonal tributaries that are sources of hippo and livestock subsidies during the wet season only. The arrows for livestock and hippo inputs do not represent the actual input values.

149

Table 6.6. Modal percentage contribution and 95% credibility intervals (in parentheses) of Bayesian isotope mixing model (SIAR) results for the contribution of different producers to different consumer trophic guilds collected during the dry and wet seasons at different sites in the Mara River basin, Kenya. SD δ^{13}C and SD δ^{15}N are the residual error, i.e., the variability in δ^{13}C/ δ^{15}N (in‰) that cannot be explained by diet alone. Models were based on both δ^{13}C and δ^{15}N. Results without residual errors were run on the SIARSOLO mode. C3, C3 producers; C4, C4 producers.

Producer/ consumer	Ngetuny Dry	Ngetuny Wet	Tenwek Dry	Tenwek Wet	Olbutyo Dry	Olbutyo Wet
C3	34 (6 - 57)	42 (6 - 74)	11 (0 - 25)	36 (1 - 52)	31 (8 - 48)	
C4	13 (1 - 23)	9 (0 - 44)	29 (16 - 41)	34 (6 - 60)	36 (15 - 54)	
Periphyton	48 (25 - 86)	43 (3 - 76)	29 (5 - 51)	36 (1 - 72)	34 (1 - 70)	
Lichens			34 (8 - 53)			
SD δ^{13}C (‰)	0.4 (0.0 - 3.2)	1.0 (0.0 - 26.8)	0.1 (0.0 - 1.2)	1.8 (0.0 - 15.7)	0.2 (0.0 - 3.0)	
SD δ^{15}N (‰)	0.2 (0.0 - 1.9)	0.9 (0.0 - 33.4)	0.2 (0.0 - 1.8)	3.0 (0.0 - 17.5)	1.2 (0.0 - 6.1)	
Invertebrate gatherers						
C3		38 (3 - 67)		38 (0 - 64)	36 (3 - 53)	
C4		14 (0 - 46)		33 (0 - 56)	36 (8 - 59)	
Periphyton		42 (5 - 79)		37 (0 - 72)	35 (0 - 60)	
SD δ^{13}C (‰)		0.9 (0.0 - 22.3)		3.2 (0.0 - 47.2)	2.3 (0.1 - 9.4)	
SD δ^{15}N (‰)		0.6 (0.0 - 14.7)		1.9 (0.0 - 56.1)	1.3 (0.0 - 6.7)	
Predatory invertebarets						
C3		40 (0 - 69)		40 (3 - 64)	36 (11 - 58)	
C4		6 (0 - 43)		26 (0 - 51)	32 (6 - 52)	
Periphyton		43 (8 - 87)		37 (0 - 70)	34 (0 - 68)	
SD δ^{13}C (‰)		1.7 (0.0 - 16.4)		0.9 (0.0 - 20.0)		
SD δ^{15}N (‰)		0.5 (0.0 - 7.1)		2.1 (0.0 - 89.5)		
Scraper invertebrates						
C3		40 (4 - 62)		37 (2 - 58)	36 (8 - 58)	
C4		13 (1 - 26)		21 (1 - 40)	30 (4 - 48)	
Periphyton		46 (19 - 86)		39 (7 - 87)	37 (1 - 72)	
SD δ^{13}C (‰)		0.3 (0.0 - 4.5)		1.6 (0.0 - 5.0)	1.5 (0.0 - 6.8)	
SD δ^{15}N (‰)		0.3 (0.0 - 3.7)		0.7 (0.0 - 2.9)	0.4 (0.0 - 4.7)	
Shredder invertebrates						
C3		6 (0 - 43)		47 (23 - 82)		
C4		23 (10 - 37)		6 (0 - 28)		
Periphyton		63 (27 - 86)		41 (1 - 66)		
SD δ^{13}C (‰)		2.1 (0.9 - 4.9)				
SD δ^{15}N (‰)		0.2 (0.0 - 2.1)				
Insectivorous fishes						
C3	36 (3 - 63)	47 (35 - 64)	3 (0 - 18)	23 (12 - 35)	24 (11 - 35)	8 (0 - 22)
C4	26 (6 - 42)	10 (0 - 46)	11 (2 - 20)	33 (23 - 40)	27 (14 - 38)	48 (44 - 52)
Periphyton	37 (4 - 74)	38 (1 - 53)	41 (20 - 57)	46 (27 - 64)	49 (28 - 75)	45 (27 - 54)
Lichens			44 (23 - 62)			
SD δ^{13}C (‰)			0.1 (0.0 - 0.8)	0.1 (0.0 - 0.8)	0.1 (0.0 - 0.7)	0.2 (0.0 - 0.9)
SD δ^{15}N (‰)			0.4 (0.0 - 1.0)	0.1 (0.0 - 0.8)	0.1 (0.0 - 1.0)	0.1 (0.0 - 0.7)
Omnivorous fishes						
C3						
C4						
Periphyton						
SD δ^{13}C (‰)						
SD δ^{15}N (‰)						

	Issey Dry	Issey Wet	Kapkimolwa Dry	Kapkimolwa Wet	Mara Main Dry	Mara Main Wet
C3	2 (0 - 11)	22 (5 - 41)	12 (2 - 37)	32 (3 - 51)	14 (0 - 29)	2 (0 - 23)
C4	12 (5 - 21)	20 (5 - 33)	17 (3 - 36)	35 (12 - 54)	43 (11 - 75)	63 (39 - 89)
Periphyton	84 (70 - 93)	58 (30 - 85)	76 (28 - 92)	36 (1 - 71)	43 (3 - 76)	30 (0 - 53)
Lichens						
SD δ^{13}C (‰)	0.6 (0.0 - 1.7)	0.7 (0.0 - 2.9)	0.1 (0.0 - 1.3)			
SD δ^{15}N (‰)	0.1 (0.0 - 1.0)	0.2 (0.0 - 2.2)	0.3 (0.0 - 2.0)			
Invertebrate gatherers						
C3	5 (0 - 25)	11 (0 - 36)	15 (0 - 35)	23 (0 - 45)		
C4	30 (15 - 44)	6 (0 - 30)	39 (16 - 58)	40 (13 - 60)		
Periphyton	67 (33 - 83)	81 (40 - 97)	42 (9 - 80)	39 (2 - 74)		
SD δ^{13}C (‰)	0.8 (0.0 - 2.7)	1.8 (0.0 - 5.0)	0.6 (0.0 - 3.6)	1.4 (0.0 - 9.5)		
SD δ^{15}N (‰)	0.2 (0.0 - 2.1)	0.6 (0.0 - 3.2)	0.3 (0.0 - 3.5)	0.8 (0.0 - 8.3)		
Predatory invertebarets						
C3	39 (2 - 64)	19 (5 - 40)	34 (5 - 53)		36 (3 - 54)	5 (0 - 50)
C4	15 (0 - 49)	19 (6 - 32)	31 (5 - 50)		35 (2 - 59)	41 (5 - 81)
Periphyton	41 (3 - 80)	63 (31 - 85)	35 (3 - 78)		35 (1 - 67)	36 (0 - 66)

Table 6.6 Continued

SD δ^{13}C (‰)	0.9 (0.0 - 22.8)	0.6 (0.0 - 2.3)	0.4 (0.0 - 6.6)		1.8 (0.0 - 33.4)	2.2 (0.0 - 40.8)
SD δ^{15}N (‰)	1.5 (0.0 - 24.8)	0.3 (0.0 - 2.3)	1.1 (0.0 - 9.5)		0.8 (0.0 - 18.3)	0.7 (0.0 - 17.3)
Scraper invertebrates						
C3	39 (3 - 61)	29 (3 - 53)	24 (3 - 43)	35 (6 - 55)	17 (0 - 47)	10 (0 - 47)
C4	28 (0 - 52)	40 (11 - 12)	28 (3 - 46)	31 (8 - 50)	38 (4 - 67)	44 (8 - 73)
Periphyton	36 (1 - 73)	36 (0 - 59)	39 (16 - 88)	37 (1 - 73)	40 (2 - 74)	41 (1 - 67)
SD δ^{13}C (‰)	1.0 (0.0 - 30.5)	0.8 (0.0 - 20.0)	1.3 (0.0 - 4.8)		1.0 (0.0 - 34.4)	1.0 (0.0 - 35.5)
SD δ^{15}N (‰)	1.8 (0.0 - 39.9)	0.7 (0.0 - 17.2)	0.2 (0.0 - 2.8)		0.7 (0.0 - 22.5)	0.7 (0.0 - 19.4)
Shredder invertebrates						
C3	38 (13 - 66)	41 (21 - 63)				
C4	22 (2 - 40)	19 (2 - 34)				
Periphyton	39 (2 - 72)	39 (8 - 68)				
SD δ^{13}C (‰)		0.4 (0.0 - 4.4)				
SD δ^{15}N (‰)		0.3 (0.0 - 4.3)				
Insectivorous fishes						
C3	15 (9 - 22)	27 (16 - 37)	26 (17 - 37)	25 (5 - 38)	13 (2 - 26)	2 (0 - 18)
C4	22 (17 - 27)	38 (30 - 45)	28 (17 - 38)	34 (20 - 44)	14 (1 - 37)	70 (57 - 82)
Periphyton	62 (52 - 73)	37 (20 - 52)	44 (27 - 64)	42 (19 - 73)	72 (40 - 94)	21 (1 - 41)
Lichens						
SD δ^{13}C (‰)	0.1 (0.0 - 0.5)	0.1 (0.0 - 1.0)	0.1 (0.0 - 0.9)	0.2 (0.0 - 1.2)	2.4 (1.4 - 3.8)	0.8 (0.0 - 1.5)
SD δ^{15}N (‰)	0.1 (0.0 - 0.6)	0.1 (0.0 - 1.1)	0.1 (0.0 - 0.8)	0.2 (0.0 - 1.7)	0.1 (0.0 - 1.0)	1.5 (0.0 - 2.6)
Omnivorous fishes						
C3			19 (9 - 34)	21 (6 - 37)	11 (4 - 20)	2 (0 - 15)
C4			20 (6 - 36)	31 (18 - 42)	24 (10 - 42)	62 (51 - 77)
Periphyton			61 (34 - 81)	51 (24 - 74)	64 (40 - 84)	36 (9 - 48)
SD δ^{13}C (‰)			1.5 (0.0 - 2.8)	0.1 (0.0 - 1.2)	0.9 (0.0 - 1.6)	0.2 (0.0 - 1.3)
SD δ^{15}N (‰)			0.1 (0.0 - 0.9)	0.1 (0.0 - 1.2)	0.1 (0.0 - 0.7)	0.4 (0.0 - 1.7)

In the Mara River, a seasonal deviation of ~2‰ in δ^{13}C of FBOM and seston was observed with lower values in the wet season (Figures 6.4). Basal resources in small agricultural streams under direct influence of livestock become ^{13}C-enriched during the dry season as a result of deposition of C4 excreta. During the wet season, low δ^{13}C values are linked to surface run-off of soil OM from the catchments that are C3 dominated (Plates 5.1 and 6.2), which is in agreement with other studies in the tropics (Bird et al., 1998). Because of increased turbidity arising from soil erosion and input of terrestrial organic matter during the wet season (Defersha and Melesse, 2012), autochthonous production is reduced making terrestrial inputs the dominant contributions to consumers. Increased turbidity can limit autochthonous production enhancing reliance of aquatic consumers on terrestrial subsidies (Mead and Wiegner, 2010; Abrantes et al., 2013). The fact that C4 sources were more important than C3 sources for most consumer groups at agriculture sites (Issey, Tenwek, Olbutyo) during the dry season (Table 6.6) is consistent with the importance of herbivore-mediated inputs. Without the herbivore inputs, I would expect to see more C3 inputs during the dry season, since surface runoff is reduced, and the proportional contribution of riparian vegetation (river zone contains more C3 than at further distance from the river course) would increase (Bird et al., 1998; Marwick et al., 2014b). At the Mara Main site, patterns are opposite to what we see at the agriculture sites (Table 6.5). During the dry season it would be expected that inputs mediated by hippos, which are constant

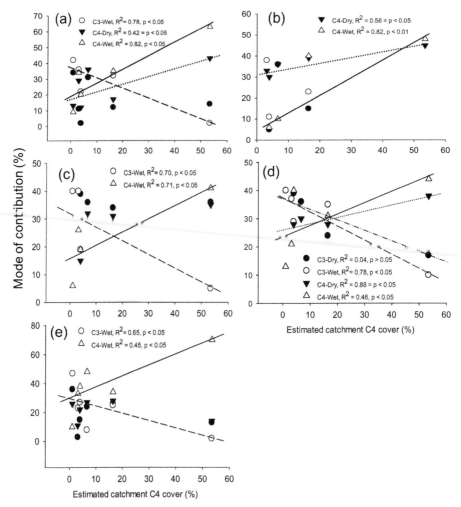

Figure 6.11. Simple linear regression (SLR) relationships between the estimated proportion of C4 vegetation cover in the catchments and the modal contribution (based on Bayesian mixing models) of C4 and C3 producers during the dry and wet seasons for (a) collector-filterers, (b) collector-gatherers, (c) predatory macroinvertebrates, (d) macroinvertebrate scrapers, and (e) insectivorous fishes. Significant relationships are indicated; full line = C4 wet season, short dotted line = C4 dry season, long dotted line = C3 wet season, and short and long dotted line = C3 dry season- only in (d).

throughout the year, would be most important relative to the wet season when material delivered from the catchment upstream are significant. However, there are indications to suggest that increased turbidity through scouring and mobilization of hippo excreta in pools by elevated discharge levels during the wet season (Plate 6.2) would limit primary production and shift the food webs to rely more on hippo subsidies. During the dry season, flow levels significantly drop in the river (McClain et al., 2014), implying that hippo faeces accumulate in pools. The slow movement of water in backwaters and side channels and the elevated nutrients brought in by hippos favour in-stream primary production which is then incorporated into the food webs (Plate 6.2, Figure 6.10). The higher $\delta^{13}C$ values for seston during the dry season which are opposite of what is seen in higher trophic levels may be explained by the mature fishes analyzed which take longer for their tissues to reflect a change in diet. Moreover, some of these species, such as *Labeo Victorianus*, *Barbus altianalis* and *Clarias gariepinus*, are migratory (Whitehead et al., 1959; Manyala et al., 2005) and the captured individuals would have been new arrivals at the Mara Main site.

Trophic positions of fishes displayed a narrow range (4.5 and 5.5) without a clear pattern among sites and fish species (Figure 6.8). This can be explained by widespread omnivory whereby most species in the Lake Victoria basin have been observed to feed on a diet mainly composed of insects and amorphous detritus (Corbet, 1961; Raburu and Masese, 2012). Lack of variability in TrPos of individual fish species among sites is indicative of fidelity to particular diets and a lack of differences in prey assemblages among sites (Beaudoin et al., 1999). Most herbivorous macroinvertebrates had TrPos between 2 and 3 suggesting mixed carbon sources. Species in these guilds consume large amounts of detritus but also assimilate significant amounts of animal material (Mantel et al., 2004, Masese et al., 2014a). The higher TrPos for some groups such as scrapers and gatherers at the agricultural sites (Figure 6.8) are indicative of nitrogen inputs from agriculture and animal activities (Plate 6.2) which were likely incorporated into the food webs (Anderson and Cabana, 2005).

So as not to bias interpretations on the important role of herbivores as vectors of delivery of C4 OM from catchments into river reaches, it is necessary to cater for possible contributions from autochthonous sources and aquatic macrophytes, which can contribute significantly to riverine OM pools (Lewis et al., 2001). First, autochthonous primary production during the dry season was low at most of the sampled reaches (mg POC: mg Chlorophyll *a* > 100) suggesting a predominance of terrestrially-derived POC (Abril et al., 2002; Finlay and Kendall, 2007); wet season primary production was expected to be lower because of turbidity. Secondly, the sampled river reaches were devoid of macrophytes which would have been an alternative source of C_3 carbon (Gichuki et al., 2001b). Thirdly, the $\delta^{13}C$ values of periphyton (mean±SD range, -25.8±1.8 to -17.3±0.9) and filamentous algae (-36.0 to -17.8 with a single sample with a value of -12.5) were low during the dry season, wet season values were lower (Figure 6.4), and in the range of the C_3 photosynthetic pathway. This implies that any autochthonous sources would most likely have contributed to the C3 OM pool in the river, and not to the C4 OM pool.

The importance of autochthonous production and terrestrial C3 producers in the 4th order Ngetuny site and the importance of periphyton in the middle reaches of the Mara mainstem during the dry season agrees with the RCC and RPM (Vannote et al., 1980; Thorp and Delong, 2002). However, the findings that terrestrial C4 sources transported into the river by large herbivores can be important for river food webs has not been captured by any model. The entire gradient of the studied river is incised with minimal interaction with the floodplain, making the possibility of floodplain-derived terrestrial vegetation to support food webs (Junk et al., 1989) unlikely. However, the importance of terrestrial C4 sources during the wet season is also related to the flushing of organic matter and nutrients from savanna grasslands into the river by surface runoff (Augustine et al., 2003; Strauch, 2011). These results concur with other studies where terrestrial sources have been found to be spatially and seasonally important for riverine food webs (Zeug and Winemiller, 2008; Hunt et al., 2012; Pingram et al., 2014).

Different trophic guilds and FFGs of invertebrates relied on different sources of energy to different extents, indicating significant spatial and taxonomic variation in sources of nutrition (c.f. Zeug and Winemiller, 2008; Pingram et al., 2014). Macroinvertebrate filterers that feed on the water column were good indicators of materials transported from upstream and reflected the dominant terrestrial energy sources at the vicinity of the sites (either C3 or C4). On the other hand, macroinvertebrate gatherers and insectivorous fishes captured the immediate influences at the vicinity of the sampled sites during the dry season. However, the findings that periphyton was the dominant source for shredders at the Ngetuny forest site are unexpected but can be explained by the fact that most of these shredders were omnivorous crabs (*Potamonautes* spp.) which, in addition to leaf litter, also feeds on other invertebrates and periphyton (Lancaster et al., 2008; Masese et al., 2014a).

Losses of mega-herbivore species from major land masses worldwide have had a significant influence on vegetation patterns and organic matter dynamics, nutrient distributions and ecosystem functioning (Zimov, 2005; Wardle et al., 2011). In Africa, large populations of savanna herbivores have been decimated and replaced by exotic livestock, which now make up more than 90 percent of large mammalian biomass of East and southern Africa (Prins, 2000; Ogutu et al., 2011). This loss of herbivores makes it difficult to understand pre-development ecosystem dynamics and establish terrestrial-aquatic food web linkages, especially those that are mediated by indigenous large animal vectors. It has been predicted for African savannas that a substantial reduction in large herbivore diversity will result in significant changes in ecosystem structure and function as well as a cascading decline in terrestrial savanna biodiversity (du Toit and Cumming, 1999). Along river valleys, reciprocal flows of subsidies by animals, or through their game paths through which materials flow from terrestrial landscapes to rivers, have also been reported (du Toit et al., 1990; Naiman et al., 2003). The large populations of herbivores in African savanna systems, such as the Mara-Serengeti ecosystem in east Africa, offer opportunities to study and infer the role of large herbivores on riverine ecosystem functioning (Naiman and Rodgers, 1997; Jacobs et al., 2007; Strauch, 2011). Thus, this study offers a glimpse into the past and, at the same time, presents evidence for increasing anthropogenic influence on riverine ecosystems structure and function.

Most rivers draining into Lake Victoria have been cleared of hippo populations, which are currently confined to rivermouths of major rivers and littoral areas around the lake (Chapter 3). As evidenced in this study, replacing hippo populations by livestock affect OM dynamics in the rivers. For instance, as ruminants, cattle rework their ingested food when chewing the cud resulting in a more refined and homogenous excreta while hippos excrete semi-digested material. The different conditioning of ingested

154

organic matter by these two herbivores likely influences nutrient cycling and ecosystem dynamics, but comparative studies are limited. In the Mara River, hippos have been linked to increased primary and secondary production (Gereta and Wolanski, 1998). At the same time the high loads of OM into the river are suspected to be responsible for fish kills during rain events after prolonged dry conditions (Amanda Subalusky, Yale University, pers comm.).

Conclusions

Partitioning the relative importance of different sources of energy supporting river food webs is critical for their management and restoration (Winemiller, 2004; Naiman et al., 2012). This study highlights the importance of considering seasonality in riverine food web studies and the important role that terrestrial herbivores play as vectors that enhance energetic terrestrial-aquatic linkages in riverine ecosystems in African savanna landscapes. Results show that different terrestrial and autochthonous sources fuel aquatic food webs, and that these resources are spatially and seasonally variable (Figure 6.10). The importance of different sources for consumer groups depends on the nature of herbivore and human influences, season and location on the fluvial continuum, and highlights the need to further examine the interaction among discharge variation, animal-mediated subsidies and taxonomic diversity for the preservation of key ecological functions in rivers of the region (Poff et al., 1997).

Chapter

7

Synthesis and conclusions

1. Overview

This dissertation is a contribution to the theory of river functioning and has improved understanding of the functioning of African tropical streams and savanna rivers. By identifying 19 shredder taxa in the study area, the highest number so far in the African tropics, I refute preexisting findings that shredder diversity is limited in African tropical streams; this number (19) compares well with other tropical and temperate regions where shredders are not limited (Table 3.5). The use of leaf litter decomposition experiments in this study has also helped highlight the important role played by shredders in organic matter processing in forested African tropical streams. This study also fills a knowledge gap by classifying 109 macroinvertebrate taxa in African tropical streams into functional feeding groups. The findings of this study also contributes to the growing evidence that shows that diet-tissue fractions of stable nitrogen ($\Delta^{15}N$) and carbon ($\Delta^{13}C$) isotopes for consumers in tropical rivers are at the low end of the range of values reported from temperate systems (e.g., Bunn et al., 2013). The use of suitable $\Delta^{15}N$ and carbon $\Delta^{13}C$ for consumers in tropical rivers will lead to more satisfactory outputs with isotope mixing models.

The documented role of large herbivores as vectors enhancing terrestrial-aquatic energetic linkages in savanna landscapes via their transfer of allochthonous C4 grasses adds to the theory of riverine ecosystem functioning. Prior to these findings C4 grasses were considered to be a poor quality resource that contributed minimally to food webs as compared with C3 vegetation (input mainly through litterfall) and periphyton (Roach, 2013 and references therein). This study also show that the relative importance of allochthonous and autochthonous sources of energy for riverine consumers are temporally variable, which is in line with the importance of seasonality and discharge variation as drivers of ecosystem functioning of tropical rivers. Many studies that consider autochthonous production to be the major source of energy for riverine food webs were conducted during dry season (Thorp and Delong, 2002; Zeug and Winemiller, 2008). Moreover, the findings of this dissertation show that rural land use activities both at the catchment- and at the reach-scale influence ecosystem functioning by modifying organic matter dynamics, light regimes, ecosystem metabolism and the distribution of macroinvretebrate taxa, mainly shredders. In this regard, this study provides information that is useful for defining future research needs and actions for sustainable management of rural agriculturally influenced streams, including their riparian zones and biological communities.

2. Synthesis

2.1 Anthropogenic influence on ecosystem functioning

As reviewed in chapter 2, changes in land cover and land use in catchment areas have modified the composition and amount of material load in streams and rivers in the Lake Victoria basin (LVB). These changes have had significant implications for water quality, organic matter (OM) processing and energy sources. The land use and land cover changes that have occurred in the upper reaches of many rivers in the LVB include clearance of indigenous vegetation and its replacement with exotic plantation species such as *Pinus* spp., *Cupressus* sp. and *Eucalyptus* spp. Leaf litter from some exotics, such as *Eucalyptus* spp., are of poorer quality than the indigenous tree species they replace (Chapter 4). Because native trees have a different regime in the shedding of their leaves (Magana, 2001), their removal and replacement with exotic species has the potential to modify the amount and timing of leaf litter input to streams. The clearing of riparian vegetation along low order streams has also created favourable light conditions for primary production and has turned many potentially hetetrophic streams autotrophic (Chapter 5).

There was support and concurrence of results and conclusions from the different methods and approaches used to measure the effects of catchment land use and riparian disturbances (animal watering, bathing and laundry washing by people) on ecosystem functioning. The macroinvertebrate functional feeding groups (FFGs) displayed a shift in composition from forest to agriculture streams that reflects a change from reliance on allochthonous leaf litter (shredder dominance) to reliance on autochthonous instream production (dominance of scrapers). Similarly, both gross primary production (GPP) and ecosystem respiration (ER) rates displayed a shift with land use change with forest streams recording higher rates of ER (GPP/ER <0.5) indicating that they were more heterotrophic than most agriculture streams (GPP/ER >0.5). Optical properties (absorbance and fluorescence based) of DOM largely indicated a predominance of terrestrially-derived DOM in forest streams and autochthonous DOM in agriculture streams, even though soil erosion from agricultural lands was suspected to be a source of aromatic DOM in agriculture streams during the wet season. These land use influences on ecosystem functioning were also tracked by results of $\delta^{13}C$ and $\delta^{15}N$ of energy sources and consumers, with a predominance of terrestrially-derived C3 vegetation at the forest sites and benthic algae (periphyton) dominated in agricultural streams during the dry season, even though terrestrial subsidies of organic matter by mammalian herbivores also contributed to food webs in agriculture streams.

2.2 Role of shredders in organic matter processing

One of the least understood attributes of African tropical streams is the diversity and abundance of shredders and their role in leaf litter processing. The few studies that have examined the functional diversity and community structure of macroinvertebrate assemblages have concluded that while macroinvertebrates are diverse and abundant, shredders are limited (Dobson et al., 2002; Masese et al., 2009b). However, these studies also indicate that identification of shredders is hampered by a lack of identification keys and the few studies that have been done have used keys developed for temperate stream fauna (e.g., Merritt et al., 2008). In some cases, these keys have been found to be misleading for some taxa in the tropics (Dobson et al., 2002; Cheshire et al., 2005), necessitating the need for development of keys and guides for Aftrican tropical stream fauna. Aided by the analysis of gut contents, I indentified a total of 19 shredders in this study and seven shredder taxa had restricted distribution to forested streams (Chapter 3). Contrary to previous findings in African tropical streams (Tumwesigye et al., 2000; Dobson et al., 2002; Abdallah et al., 2004) my findings show that shredders are important components of upland streams but their distribution is limited by changes in water quality and riparian alterations by human and livestock activities.

The consequence of limited shredder distribution was highlighted by their role in leaf litter processing in the streams (Chapter 4). In addition to highlighting the applicability of leaf litter processing and the composition of invertebrate FFGs as functional and structural indicators, respectively, of ecological health, Chapter 4 also highlights interactions among catchment land use, riparian activities and seasonality as drivers of ecosystem functioning in upland tropical streams. While catchment land use is an important determinant of temperature and litter biomass in the studied streams, reach-scale (riparian) influences which determine leaf litter quality through exotic introductions and reduced water quality were also important in structuring macroinvertebrate communities and determing their functional organization, with effects propagating to the processing of leaf litter. The findings of this dissertation (Chapters 3 and 4) are significant contributions to the growing data on the functional organization of tropical streams, including dietary requirements and trophic relationships. The findings will likely spur further research in these topics in the African tropics. By using the ratios of functional feeding groups as surrogates of ecosystem attributes (*sensu* Merritt et al., 2002) and functional indicators of ecological health, this study will also contribute to better assessment of human influences and management of upland African tropical streams that are threatened by land cover and land use changes (e.g., Jinggut et al., 2012).

2.3 Land use effect on DOM composition and ecosystem metabolism

Catchment land use change and the loss of riparian corridors to deforestation have had a disproportiate influence on the functioning of headwater streams mainly through changes in biological communities, organic matter dynamics, nutrient input and in-stream primary production (Bilby and Bisson, 1992). With the realization that the quality of dissolved organic matter (DOM), in addition to its quantity, is important in the understanding of carbon dynamics in streams (Wetzel, 1992; Battin et al., 2008), a number of studies have sought to link DOM composition in streams with land use in the catchments they drain (Wilson and Xenopoulos, 2009; Graeber et al., 2012). However, much of our understanding of the effects of land use change on organic matter dynamics in streams and rivers is largely based on studies of temperate ecosystems (Graeber et al., 2012; Wilson and Xenopoulos, 2009; Staehr et al., 2012). This is disproportionate to the important role played by tropical streams and rivers that transport >60% of the global riverine carbon (Ludwig et al., 1996; Schlünz and Schneider, 2000). Moreover, while findings on the influence of agricultual land use on DOM composition in streams have been unequivocal, the responses in specific properties of DOM have been variable. For instance increased contributions of microbially derived and structurally less complex DOM to the DOM pool in agricture streams have been reported (Wilson and Xenopoulos, 2009; Williams et al., 2010). In contrast, Graeber et al. (2012) found the percentage of land under agricultural use to increase the amount of structurally complex and aromatic DOM in streams. These apparently divergent findings suggests that the proportion of agricultural land use within a catchemnt alone is unlikely to explain patterns in DOM composition in recipient streams. Land use history, soil type, tillage practice or technique, catchment topology, climate and alterations to hydrological residence times and flow paths, in addition to the uptake rates and provenance to metabolism of the different DOM pools, have all been found to contribute somewhat to DOM concentration and composition in streams (Ogle et al., 2005; Ewing et al., 2006). The implications of these findings are such that in order to improve our understanding of models of the global carbon cycle, more data are needed on headwater streams across different biomes and climates (Battin et al., 2008). This is more so the case in Afropical streams which are underrepresented in global models of riverine carbon cycling. More studies that examine the spatial and temporal variation in DOM amount and composition in agricultural streams are also needed to help reconcile disparate responses that have been reported for specific attributes of DOM in the literature.

I compared the spatial and temporal variation in DOM amount and composition in headwater streams in the Mara River basin, Kenya (Chapter 5). To allow better understanding of sources of DOM and its cycling in tropical streams, the longitudinal variation in the composition of DOM was investigated as streams increase in size (Vannote et al., 1980) and relationships among DOC concentration, DOM composition and ecosystem metabolism in the streams. In the upland streams of the Mara basin, forested streams exported more structurally complex and high molecular weight DOM than streams draining agricultural land. However, aromaticity of DOM was high even in agriculture streams during the wet season, and this was attributed to the mobilization of the upper soil layers which are rich in soil organic matter during tillage and subsequent transport into streams by surface runoff. Evidence of elevated levels of sediments washed by surface runoff from farmlands, footpaths by people and livestock, and unpaved roads were observable in agriculture streams during and after rain events (Plate 5.1). Measures of ecosystem metabolism responded to the land use change with higher rates of GPP and ER in agriculture streams. DOM composition seemed to track this variability whereby forest streams were net heterotrophic because of the high levels of canopy cover while many of the agriculture streams were net autotrophic as a result of high rates of primary production favoured by higher nutrient levels (mainly dissolved fractions of nitrogen) and open canopy.

Seasonality was important and through its control on material loads in streams and rivers, longitudinal and lateral hydrologic linkages, dry- and wet-season differences in rates of GPP and ER were observed. Increased rates of GPP were recorded during the dry compared with the wet season. Higher rates were also recorded at forest streams during the dry season, though this was not significantly different from dry season metabolism at agriculture streams, suggesting that level of canopy cover was also dependent on seasonality. During the dry season shedding of leaves was higher (Magana, 2001), and some tree species shed all their leaves (pers. obser.). In the study area, the effect of this phenomenon was a slightly open canopy in forest streams during the dry season; it is much easier to walk through the forest and along streams for the same reason, as opposed to during the wet season when the vegetation is denser and undergrowth more impenetrable. During this time, the levels of insolation are higher and for longer periods due to limited cloud cover and this likely contributed to higher GPP rates in forest streams during the dry season.

Analysis of DOM composition on the longitudinal gradient captured shifts from allochthonous to autochthonous sources, and this agrees with the river continuum concept (Vannote et al., 1980). Stream size mainly influenced the molecular size of DOM through photodegradation as streams became wider and insolation increased. However, the longitudinal gradients also mimicked the land use gradient making it difficult to reconcile patterns in some properties of DOM, such as reducing SUVA254 and S_R with increasing stream size.

My findings help elevate the role of land use, and specifically agriculture, as a source of structurally altered DOM to surface waters. Potential differences in catchment characteristics, and for my case land use history and usage and the techniques - plough and hoe vs intensive mechanization- need to be addressed in future studies to better understand the effect of different agricultural practices and intensities on DOM dynamics in aquatic ecosystems. Measures of ecosystem metabolism in these streams help complement the role of tropical streams and rivers as important components of the global carbon cycle and add to the growing understanding of the effects of agricultural land use on riverine ecosystem functioning.

2.4 Energy sources for riverine food webs

In the ongoing discussions, comparisons have been made between temperate and tropical streams and rivers in terms of organic matter processing, food web attributes and the major sources of energy fuelling metazoan food webs. Many of the studies in tropical headwater streams have supported the importance of autotrophic production over terrestrial inputs as a major source of energy for aquatic consumers (Mantel et al., 2004; Li and Dudgeon, 2008; Coat et al., 2009). However, even with these emerging shared characteristics, there is a general consensus on the high heterogeneity in the structure and functioning of tropical riverine ecosystems (Boulton et al., 2008) and the influence of discharge variation has not been extensively explored. Highland montane streams are likely to be functionally different from their lowland savanna counterparts. In the African savannas, large mammalian herbivores were once key features of these landscapes (Prins, 2000; Ogutu et al., 2011) with the potential to influence streams and rivers through vectoring of terrestrial organic matter and nutrients into aquatic environments when watering. However, most of the herbivores have been lost as a result of land use change and have been replaced in large areas by livestock. The implications of these losses on energetic terrestrial-aquatic linkages are hardly understood. Lack of comparable studies across regions and systems has limited generalizations about determinants of ecosystem structure and function and predicting the likely impacts of global and regional human disturbances, notably land use change and the associated alterations of the natural flow regimes of streams and rivers. Partitioning the relative importance of different sources of energy supporting river food webs is critical for their management and restoration (Winemiller, 2004; Naiman et al., 2012). In this dissertation I pursued these questions with the aim of understanding the influence of land use change and mammalian herbivores on the relative importance of different sources of energy for consumers in the Mara River (Chapter 6).

The SIAR model results in this dissertation (Chapter 6) show that overally, periphyton is the major source of energy for consumers in the Mara River during the dry season. This is in agreement with many dry season studies that have shown periphyton or algae to be the predominant source of energy for riverine consumers (Thorp and Delong, 1994, 2002; Delong and Thorp 2006; Roach, 2013). During the wet season, however, allochthonous sources gained importance and at sites influenced by mammalian herbivores (livestock, ungulates and hippos) the importance of C4 producers over C3 producers increased (Figure 6.10). I show that both terrestrial and autochthonous sources fuel aquatic food webs and that for consumers in rivers that drain savanna landscapes these sources vary seasonally with seasonal pulses of flow and riparian inputs, and spatially in relation to land use influences and mammalian herbivore inputs (Figure 6.10). However, one of the intriguing findings in this regard is the importance of autochthonous sources for consumers in river reaches clearly dominated by hippo inputs during the dry season. During this period, it is expected that because of limited upstream sources and lack of runoff from the catchments, terrestrial inputs by hippos would dominate in-stream secondary production, especially considering that bioturbation by hippos would increase turbidity and subsequently reduce primary production. However, there is evidence to suggest that despite the presence of hippos in the Mara River, autochthonous production can dominate energy sources during the dry season. Pools created by hippos reduce water velocity, increase habitat complexity and create backwater habitats where primary production can occur (Thorp and Delong, 1994; Gereta and Wolanski, 1998; Mosepele et al., 2009).

2.5 Contribution to models of riverine ecosystem functioning

This dissertation provides support for some tenets of both the River Continuum Concept (Vannote et al., 1980) and Riverine Productivity Model (Thorp and Delong, 1994, 2002). In forest streams, shredders were diverse and played major roles in leaf litter processing but the dominance of upstream derived energy was not found to be important for downstream river reaches during both the dry and wet seasons. In agreement with RPM (Thorp and Delong, 2002), benthic algae (periphyton) were the dominant basal carbon resource supporting consumers in the middle reaches of the Mara River and its two main tributaries, the Nyangores and the Amala, during the dry season (Figure 6.10). While this is also in agreement with the RCC (Vannote et al., 1980), the dominance of terrestrially-derived C4 sources at herbivore influenced sites during the wet season is unprecedented and reflects the important role played by mammalian herbivores as vectors of terrestrial-aquatic energetic linkages in African savanna landscapes (Chapter 6). However, land use change and the continued loss of large mammals in savanna landscapes is likely to influence the functioning of riverine ecosystems by eliminating this energetic linkage. The livestock that are replacing wildlife herbivores need to be investigated to determine whether they have a similar influence.

3. Management concerns and recommendations

Because of its transboundary nature (Kenya/Tanzania) and by hosting the Masai Mara National Reserve (Kenya) and Serengeti National Park (Tanzania), which are internationally renowned tourist attractions, the Mara River basin has attracted a lot of attention over the last decade by both policy makers and scientists. One of the past projects is the Global Water for Sustainability (GLOWS) program that was funded by the United States Agency for International Development (USAID), which assessed the environmental flow requirements in the Mara River and funded a number of other studies that have made recommendations on water availability and use, including the impacts of land use change on water resources and ecosystem functioning (e.g., Melesse et al., 2008; Omengo, 2010; Mango et al., 2011; Defersha and Melesse, 2012; Minaya et al., 2011; McClain et al., 2014). Human activities in the Mara River basin that have implications for the functioning of the Mara River include land use change, soil erosion, increasing water demands and uncontrolled water abstractions, loss of riparian indigenous vegetation and its replacement with exotic tree species, and loss of large mammalian wildlife populations and their replacement with livestock (this study, Serneels et al., 2001; Lamprey and Reid, 2004; Derfersha and Melesse, 2012; Mati et al., 2008; Mango et al., 2011; Ogutu et al., 2011; Dessu et al., 2014). While non-point sources of pollution are difficult to prevent, adoption of best management practices (BMPs) will help improve soil and water conservation in the basin.

The current ecological status of the Mara River is bound to change because of development activities that have been lined up in the basin as part of Kenya's Vision 2030 (GoK, 2007, 2008). Currently no major reservoir exist in the study area; however two mini-sized multi-purpose dams are planned on the Mara River: Mugango on the Nyangores tributary above Tenwek which is the largest and Norera on the Amala tributary in Kenya and at Borenga in Tanzania (NBI, 2011). These dams will provide water for irrigation, domestic use, fisheries and flood control. Plans are also underway to increase the area under irrigation in the lower Nyangores River sub-basin at Chepalungu. These changes threaten to alter flow regimes and material load in ways that will likely impair ecosystem integrity and functionality of the rivers. In addition, agricultural land use is likely to increase and intensify in the coming years to meet increasing food requirements and provide income for the growing human population, and these increases are more likely to occur in the semi-arid middle reaches with more water demands for irrigation (Serneels et al., 2001; Mati et al., 2008; Kilonzo, 2014). The government of Kenya through the National Cereals and Produce Board has also encouraged fertilizer application on farms by subsidizing prices for farmers at 70% of the market price. All these changes in the basin have implications for the functioning of the river through modifications in vegetation and organic matter dynamics, run-off processes, erosion, nutrient input and flow levels.

A number of management decisions are needed, and some have been made, to safeguard the long-term sustainability of the Mara River basin. Catchment reforestation of degraded sections of the Mau Forest (GoK, 2009), preventing further loss of riparian vegetation and reforestation using indigenous tree species are some of the recommendations that will help restore the integrity of streams in agricultural areas. To achieve this, the government of Kenya has set aside resources to rehabilitate the Mau Forest (GoK, 2009). At the local scale, community forest users associations (CFAs) and water resource users associations (WRUAs) are participating in reforestation activities and rehabilitation of riparian zones. In the short term, there is a need for continued financial support and capacity building of these organizations to achieve their objectives as laid out in their management plans- limiting soil erosion, reforestation of degraded forest areas, rehabilitation and demarcation of riparian zones and monitoring water use and abstractions among other activities. In the long-term, these associations will need to be self-reliant and sustainable. However, the success of any initiative in the basin will ultimately have to address the drivers of environmental change themselves, i.e., human population growth, land use policy and the high poverty levels.

While the current water quality situation in the Mara River is not critical, there have been cases of microbial contamination during the dry season when flows in the streams and the rivers are much reduced. With increasing demands occasioned by the growing animal and human populations, water quantity and quality will become limiting in the coming years. This will be exarcerbated by inadequate sanitation facilities, improper discharge of effluents from toilets into rivers, improper use of agrochemicals, watering of livestock directly in rivers, bathing and washing of clothes in rivers, and poor solid and liquid disposal facilities (NBI, 2008). To tackle these problems, liquid and solid waste disposal should be managed well and a water quality monitoring program in Mara River basin should be established. Most importantly, adherence to the rule of law, enforcement and compliance in environmental management will be the key to the sustainability of ecosystem functions and provision of ecosystem services.

4. Future research

Stream and river studies from African biomes such as montane forests and savannas are underrepresented in the tropical literature and an understanding of the structure and function in these ecosystems is generally missing in assessment of emerging trends (Boyero et al., 2009; Dudgeon et al., 2010). The climatic difference and seasonal heterogeneity of tropical African rivers offer an opportunity to expand and inform the existing tropical database. While this dissertation has addressed the role of shredders in leaf litter processing in upland montane streams, there is a need to extend the study by comparing streams in different climates (savanna vs montane) and altitudes (e.g., Boyero et al., 2009).

Future research should quantify the contributions of nutrients by potamodromous fish species to riverine food webs (e.g., Walters et al., 2009), especially in the lower reaches of rivers draining into Lake Victoria. The spatial isotopic variation of some migratory fish species such as *Labeo victorianus*, *Barbus altianalis* and *Clarias gariepinus* (Ojwang et al., 2007) has the potential to allow the quantification of movements between habitats (Rasmussen et al., 2009). Inclusion of isotopes of other elements such as sulphur and hydrogen would improve discrimination between floodplain, riverine and lake derived carbon (e.g., Jardine et al., 2011). The use of otolith microchemistry would also provide another approach to identify and determine fish movements between habitats (Campana et al., 2005; Elsdon, et al., 2008; Smith and Kwak, 2014) and migrations by potamodromous fishes from the lake into influent rivers. Using isotopes and dietary analysis to identify ontogenetic shifts in resource use, trophic position and niche width within species and overlaps among species (Ojwang et al., 2004; Layman et al., 2007; Davis et al., 2012; Jackson et al., 2012), could be important for describing food web attributes and linkages, with potential implications for species management and fisheries productivity.

Terrestrial invertebrates were not sampled in this study and this could have illustrated more direct linkages between riverine food webs and terrestrial resources, especially in headwater streams where they are known to contribute significantly to fish biomass (Nakano et al., 1999; Baxter et al., 2005). Future studies should use stable isotopes to estimate turnover rates for different species and trophic groups during different flow conditions. This will strengthen interpretations of energy sources and pathways along the Mara River. This is specifically true in the middle reaches of the Mara River that are inhabited by hippos and crocodiles which makes sampling of resources and consumers very difficult and sometimes dangerous. Similarly, in order to improve estimates of source contributions to consumers, diet–tissue fractionation values of stable nitrogen and carbon isotopes will have to be established under control experiments where the isotopic values of diets are well known. This will help minimize uncertainty in mixing model outputs and the calculation of trophic levels in food web studies in tropical aquatic systems. In addition, future studies on sources and flows of energy should quantify organic-matter flows in order to determine the absolute basis of metazoan production of aquatic systems (Benke and Wallace, 1980; Cross et al., 2013), as opposed to a qualitative determination of energy sources and flow in food webs.

Given the role that large migratory herbivores play in the ecology of large rivers in the region (this dissertation), studies are needed to determine the effects of removing hippos from river reaches. Most rivers draining into Lake Victoria have been cleared of hippo populations, which are currently confined to mouths of major rivers and littoral areas around the lake (Chapter 2). As evidenced in this study, replacing hippo populations by livestock affects OM dynamics in the rivers. For instance, as ruminants, cattle rework their ingested food when chewing the cud resulting in a more refined and homogenous excreta, as opposed to hippos which excrete semi-digested material. The disparate conditioning of ingested organic matter by these two herbivores likely influences nutrient cycling and ecosystem dynamics, but comparative studies are limited. Whether the presence of hippos results in increased primary and secondary production (e.g., fisheries) is also not clear because of the contrasting outcomes that have been reported. In the Mara River, hippos have been linked to increased primary and secondary production (Gereta and Wolanski, 1998), but more data are needed to determine whether their bioturbation has negative effects on ecosystem productivity. It has been estimated that around 10% of the suspended matter in the Mara River is composed of hippo excreta (Dutton, 2012). At the same time the high loads of OM into the river are suspected to be responsible for fish kills during rain events after prolonged dry conditions, but more data are needed to answer this unequivocally. There is also a need to determine the role of livestock and hippos in the transfer of other nutrients such as N, P, and Si from land to streams and rivers since they could have a major influence on the productivity of aquatic primary producers and play a role in eutrophication of aquatic ecosystems. There is also a need to study the influence on ecosystem functioning of livestock replacing wildlife in the middle basin and in the Talek region.

While additional research in Mara has much to contribute to emerging models of tropical river ecology, the most pressing research needs in the region are, by far, those investigating the impacts of anthropogenic change on ecosystem functions and associated losses of ecosystem services used by people. Priority here should be given to research focused on the drivers of environmental change as opposed to the manifestations of the changes themselves. In this regard research is needed into the social and governance structures in the basin and their linkage to environmental change. This is line with integrated water resources management that calls for a multidisciplinary approach to the management of complex watershed resources, especially under the current times of change.

8. References

Abdallah A, De Mazancourt C, Elinge MM, Graw B, Grzesiuk M, Henson K, Kamoga M, Kolodziejska I, Kristersson M, Kuria A, Leonhartsberger P, Matemba RB, Merl M, Moss B, Minto C, Murfitt E, Musila SN, Ndayishiniye J, Nuhu D, Oduro DJ, Provvedi S, Rasoma RV, Ratsoavina F, Trevelyan R, Tumanye N, Ujoh VN, Van de Wiel G, Wagner T, Waylen K, Yonas M. 2004. Comparative studies on the structure of an upland African stream ecosystem. Freshwater Forum 21: 27–47.

Abrantes KG, Barnett A, Marwick TR, Bouillon S. 2013. Importance of terrestrial subsidies for estuarine food webs in contrasting East African catchments. Ecosphere 4: Article 14.

Abril G, Nogueira M, Etcheber H, Cabec¸adas G, Lemaire E, Brogueira MJ. 2002. Behaviour of organic carbon in nine contrasting European estuaries. Estuarine, Coastal and Shelf Science 54: 241–62.

Acuña V, Díez JR, Flores L, Meleason M, Elosegi A. 2013. Does it make economic sense to restore rivers for their ecosystem services? Journal of Applied Ecology doi: 10.1111/1365-2664.12107

Akiyama T, Kajumulo AA, Olsen S. 1977. Seasonal variations of plankton and physicochemical condition in Mwanza Gulf, Lake Victoria. Bullettin of the Freshwater Fisheries Research Laboratories, Tokyo 27: 49–61.

Allan JD. 2004. Landscapes and riverscapes: the influence of land use on stream ecosystems. Annual Review of Ecology, Evolution and Systematics 35: 257–284.

Aloo PA. 2003. Biological Diversity of the Yala swamp Lakes, with special emphasis on fish species composition, in relation to changes in the Lake Victoria Basin (Kenya): threats and conservation measures. *Biodiversity and conservation* 12: 905-920.

Anderson C, Cabana G. 2005. $\delta^{15}N$ in riverine food webs: effects of N inputs from agricultural watersheds Canadian Journal of Fisheries and Aquatic Sciences 62: 333-340.

Anesio AM, Theil-Nielsen J, Graneli W. 2000. Bacterial growth on photochemically transformed leachates from aquatic and terrestrial primary producers. Microbial Ecology 40: 200–208.

Angradi TR. 1996. Inter-habitat variation in benthic community structure, function, and organic matter storage in 3 Appalachian headwater streams. Journal of the North American Benthological Society 15: 42–63.

Anyah RO, Semazzi FHM. 2007. Variability of East African rainfall based on multiyear RegCM3 simulations. International Journal of Climatology 27: 357–71.

APHA (American Public Health Association). 1998. Standard Methods for the Examination of Water and Wastewater, 20th edition. American Public Health Association, American Water Works Association, Water Environment Federation, Washing-ton, DC.

Ardón NM, Pringle CM. 2008. Do secondary compounds inhibit microbial- and insect-mediated leaf breakdown in a tropical rain forest stream, Costa Rica? Oecologia 155: 311–323.

Arimoro FO, Obi-Iyeke GE, Obukeni PJO. 2012. Spatiotemporal variation of macroinvertebrates in relation to canopy cover and other environmental factors in Eriora River, Niger Delta, Nigeria. Environmental Monitoring and Assessment 184: 6449–6461.

Arthington AH, Naiman RJ, McClain ME, Nilsson C. 2010. Preserving the biodiversity and ecological services of rivers: new challenges and research opportunities. Freshwater Biology 55: 1–16.

Aufdenkampe AK, Mayorga E, Raymond PA, Melack JM, Doney SC, Alin SR, Aalto RE, Yoo K. 2011. Riverine coupling of biogeochemical cycles between land, oceans, and atmosphere. Frontiers in Ecology and the Environment 9: 53-60.

Augustine DJ, McNaughton SJ, Frank DA. 2003. Feedbacks between soil nutrients and large herbivores in a managed savanna ecosystem. Ecological Applications 13: 1325–1337.

Awange JL, Sharifi MA, Ogonda G, Wickert J, Grafarend EW, Omulo MA. 2008. The falling Lake Victoria water level: GRACE TRIMM and CHAMP satellite analysis of the lake basin. Water Resources Management 22: 775–96.

Baker A, Bolton L, Newson M, Spencer R. 2008. Spectrophotometric properties of surface water dissolved organic matter in an afforested upland peat catchment. Hydrological Processes 22: 2325-2336.

Baker A, Spencer RGM. 2004. Characterization of dissolved organic matter from source to sea using fluorescence and absorbance spectroscopy. Science of The Total Environment 333: 217-232

Balirwa JS, Bugenyi FWB. 1980. Notes on the fisheries of the River Nzoia, Kenya. Biological Conservation 18: 53–58.

Balirwa JS. 1979. A contribution to the study of the food of six cyprinid fishes in three areas of the Lake Victoria basin, East Africa. Hydrobiologia 66: 65-72.

Barron C, Apostolaki ET, Duarte C. 2014. Dissolved organic carbon fluxes by seagrass meadows and macroalgal beds. Name: Frontiers in Marine Science 1, 42.

Battin TJ, Kaplan LA, Findlay S, Hopkinson CS, Marti E, Packman AI, Newbold JD, Sabater F. 2008. Biophysical controls on organic carbon fluxes in fluvial networks. Nature Geosciences 1: 95-100.

Bavor HJ, Waters MT. 2008. Buffering Performance in a Papyrus-Dominated Wetland System of the Kenyan Portion of the Lake Victoria Basin. In: J Vymaza (ed), Wastewater Treatment, Plant Dynamics and Management in Constructed and Natural Wetlands. Springer, Netherlands, 33-38pp.

Baxter CV, Fausch KD, Saunders WC. 2005. Tangled webs: reciprocal flows of invertebrate prey link stream and riparian zones. Freshwater Biology 50: 201–220.

Bayley PB. 1995. Understanding large river-floodplain ecosystems. BioScience 45: 153-158.

Beaudoin CP, Tonn WM, Prepas EE, Wassenaar LI. 1999. Individual specialization and trophic adaptability of northern pike (Esox lucius): an isotope and dietary analysis. Oecologia 120: 386–396.

Benstead JP, Pringle CM. 2004. Deforestation alters the resource base and biomass of endemic stream insects in eastern Madagascar. Freshwater Biology 49: 490–501.

Bernot MJ, Sobota DJ, Hall RO, Mulholland PJ, Dodds WK, Webster JR, Tank JL, Ashkenas LR, Cooper LW, Dahm CN, Gregory SV, Grimm NB, Hamilton SK, Johnson SL, McDowell WH, Meyer JL, Peterson B, Poole GC, Valett HM, Arango C, Beaulieu JJ, Burgin AJ, Crenshaw C, Helton AM, Johnson L, Merriam J, Niederlehner BR, O'Brien JM, Potter JD, Sheibley RW, Thomas SM, Wilson KYM. 2010. Inter-regional comparison of land-use effects on stream metabolism. Freshwater Biology 55: 1874-1890.

Bilby RE, Bisson PA. 1992. Allochthonous versus Autochthonous Organic Matter Contributions to the Trophic Support of Fish Populations in Clear-Cut and Old-Growth Forested Streams. Canadian Journal of Fisheries and Aquatic Sciences 49: 540-551.

Bilby RE, Fransen BR, Bissson PA. 1996. Incorporation of nitrogen and carbon from spawning coho salmon into the trophic system of small streams: evidence from stable isotopes. Canadian Journal of Fisheries and Aquatic Sciences 53: 164–173.

Bishai HM, Abu-Gideiri YB. 1965. Studies on the biology of the genus Synodontis at Khartoum, age and growth. Hydrobilogia 26: 58-97.

Black E. 2005. The relationship between Indian Ocean sea-surface temperature and East African rainfall. Philosophical Transactions of the Royal Society A: Mathematical, Physical and Engineering Sciences 363: 43–47.

Blackett HL. 1994. Forest inventory report no. 3: Kakamega. Kenya Indigenous Forest Conservation Programme, Nairobi, Kenya.

Bonada N, Prat N, Resh VH, Statzner B. 2006. Developments in aquatic insect biomonitoring: A comparative analysis of recent approaches. Annual Review of Entomology 51: 495–523.

Bond T, Sear D, Sykes T. 2014. Estimating the contribution of in-stream cattle faeces deposits to nutrient loading in an English Chalk stream. Agricultural Water Management 131: 156– 162.

Bond TA, Sear DA, Edwards ME. 2012. Temperature-driven river utilisation and preferential defecation by cattle in an English chalk stream. Livestock Science 146: 59–66.

Bootsma HA, Hecky RE. 1993. Conservation of the African Great Lakes: a limnological perspective. Conservation Biology 7: 644–656.

Borda JS. 2012. Effect of organic matter quality on ecosystem metabolism in the upper catchment of the Mara River, Kenya. MSc thesis, UNESCO-IHE Insitute for Water Education, Delft, The Netherlands.

Bormann FH, Likens GE. 1967. Nutrient cycling. Science 155(3761): 424-429.

Bott TL, Newbold JD, Arscott DB 2006a. Ecosystem metabolism in Piedmont streams: reach geomorphology modulates the influence of riparian vegetation. Ecosystems 9:398–421.

Bott TL, Montgomery DS, Newbold JD, Arscott DB, Dow CL, Aufdenkampe AK, Jackson JK, Kaplan LA. 2006b. Ecosystem metabolism in streams of the Catskill Mountains (Delaware and Hudson River watersheds) and Lower Hudson Valley. Journal of the North American Benthological Society 25: 1018–1044.

Bott TL, Newbold JD. 2013. Ecosystem metabolism and nutrient uptake in Peruvian headwater streams. International Review of Hydrobiology 98: 117–131.

Bouillon S, Yambélé A, Spencer RGM, Gillikin DP, Hernes PJ, Six J, Merckx R, Borges AV. 2012. Organic matter sources, fluxes and greenhouse gas exchange in the Oubangui River (Congo River basin). Biogeosciences 9: 2045–2062.

Boulton AJ, Boyero L, Covich AP, Dobson MK, Lake PS, Pearson RG, 2008. Are tropical streams ecologically different from temperate streams? In: Tropical Stream Ecology (Ed. Dudgeon D), Academic Press, San Diego (Aquatic Ecology Series): 257-284pp.

Boulton AJ, Boon PI. 1991. A review of methodology used to measure leaf litter decomposition in lotic environments: Time to turn over an old leaf. Australian Journal Marine and Freshwater Research 42: 1-43.

Boyero L, Pearson RG, Dudgeon D, Ferreira V, Graca MAS, Gessner MO, Boulton AJ, Chauvet E, Yule CM, Albarino RJ, Ramırez A, Helson JE, Callisto M, Arunachalam M, Chara J, Figueroa R, Mathooko JM, Goncalves JFJr, Moretti MS, Chara-Serna AM, Davies JN, Encalada AC, Lamothe S, Buria LM, Castela J, Cornejo A, Li AOY, M'Erimba C, Villanueva VD, Zuniga MC, Swan CM, Barmuta LA. 2011a. Global patterns of stream detritivore distribution: implications for biodiversity loss in changing climates. Global Ecology and Biogeography 21: 134–141.

Boyero L, Pearson RG, Dudgeon D, Graça MAS, Gessner MO, Albariño RJ, Ferreira V, Yule CM, Boulton AJ, Arunachalam M, Callisto M, Chauvet E, Ramírez A, Chará J, Moretti MS, Gonçalves JF, Helson JE, Chara-Serna AM, Encalada AC, Davies JN, Lamothe S, Cornejo A, Castela J, Li AOY, Buria LM, Villanueva VD, Zuniga MC, Pringle CM.. 2011b. Global distribution of a key trophic guild contrasts with common latitudinal diversity patterns. Ecology 92: 1839–1848.

Boyero L, Pearson RG, Gessner MO, Barmuta L, Ferreira V, Graça MAS, Dudgeon D, Boulton AJ, Callisto M, Chauvet E. et al. 2011c. A global experiment suggests climate warming will not accelerate litter decomposition in streams but might reduce carbon sequestration. Ecology Letters 14: 289-294.

Boyero L, Ramirez A, Dudgeon D, Pearson RG. 2009. Are tropical streams really different? Journal of the North American Benthological Society 28: 397–403.

Bray JR, Curtis JT. 1957. An ordination of the upland forest communities of southern Wisconsin. Ecological Monographs 27: 325-349.

Brito EF, Moulton TP, Souza ML, Bunn SE. 2006. Stable isotope analysis in microalgae as the predominant food source of fauna in a coastal forest stream, south-east Brazil. Austral Ecology 31: 623–633.

Brown D. 1994. Freshwater Snails of Africa and their Medical Importance. Taylor & Francis: London.

Bunn SE, Arthington AH. 2002. Basic principles and ecological consequences of altered flow regimes for aquatic biodiversity, Environmental Management 30: 492-507.

Bunn SE, Davies PM, Kellaway DM. 1997. Contribution of sugar cane and invasive pasture grass to the aquatic food web of a tropical lowland stream. Marine and Freshwater Research 48: 173–179.

Bunn SE, Davies PM, Winning M. 2003. Sources of organic carbon supporting the food web of an arid zone floodplain river. Freshwater Biology 48: 619-635.

Bunn SE, Leigh C, Jardine TD. 2013. Diet–tissue fractionation of $\delta^{15}N$ by consumers from streams and rivers. Limnology & Oceanography 58: 765–773.

Bunn SE, Davies PM, Mosisch TD. 1999. Ecosystem measures of river health and their response to riparian and catchment degradation. Freshwater Biology 41: 333–345.

Cadwalladr DA. 1965. The decline in Labeo victorianus (Boulenger), (Pisces: Cyprinidae) fishery of Lake Victoria and associated deterioration in some indigenous fishing methods in the Nzoia river, Kenya. East African Agriculture and Forestry Journal 30: 249–256.

Camacho, R., L. Boyero, A. Cornejo, A. Ibáñez, and R. G. Pearson. 2009. Local variation in shredder numbers can explain their oversight in tropical streams. Biotropica 4: 625–632.

Cammack WL, Kalff J, Prairie YT, Smith EM. 2004. Fluorescent dissolved organic matter in lakes: Relationships with heterotrophic metabolism. Limnology and Oceanography 49: 2034-2045.

Campana SE. 2005. Otolith science entering the 21st century. Marine and Freshwater Research 56: 485–495.

Campbell L, Hecky, RE, Wandera SB. 2003. Stable isotope analysis of food web structure and fish diet in Napoleon and Winam Gulfs, Lake Victoria East Africa. Journal of Great Lakes Research 29: 243–257.

Canhoto C, Laranjeira C. 2007. Leachates of Eucalyptus globulus in Intermittent Streams Affect Water Parameters and Invertebrates. International Review of Hydrobiology 92: 173-182.

Carabel S, Godinez-Dominguez E, Verisimo P, Fernandez L, Freire J. 2006. An assessment of sample processing methods for stable isotope analyses of marine food webs. Journal of Experimental Marine Biology and Ecology 336: 254-261.

Cawley KM, Campbell J, Zwilling M, Jaffé R. 2014. Evaluation of forest disturbance legacy effects on dissolved organic matter characteristics in streams at the Hubbard Brook Experimental Forest, New Hampshire. Aquatic Sciences 76: 611-622.

Cederholm JC, Kunze MD, Murota T, Sibatani A. 1999. Pacific salmon carcasses: essential contributions of nutrients and energy for aquatic and terrestrial ecosystems. Fisheries 24: 6–15.

Chapman CA, Chapman LJ. 1996. Mid-elevation forests: a history of disturbance and regeneration. Pages 385–400 in T. R. McClanahan and T. P. Young (editors). East African ecosystems and their conservation. Oxford University Press, New York.

Chapman CA, Chapman LJ. 2003. Deforestation in tropical Africa: impacts on aquatic ecosystems. Conservation, ecology, and management of African freshwaters, 229-246 pp.

Chapman LJ, Balirwa J, Bugenyi FWB, Chapman CA, Crisman TL. 2001. Wetlands of East Africa: Biodiversity, exploitation, and policy perspectives. Pp. 101-132. (In). B. Gopal (editor) Wetlands Biodiversity. Backhuys Publisher, Leiden.

Chará-Serna AM, Chará JD, Zúñiga MC, Pearson RG, Boyero L. 2012. Diets of leaf litter-associated invertebrates in three tropical streams. Annales de Limnologie 48: 139–144.

Cheshire K, Boyero L, Pearson RG. 2005. Food webs in tropical Australian streams: shredders are not scarce. Freshwater Biology 50: 748–769.

Clapcott JE, Bunn SE. 2003. Can C4 plants contribute to aquatic food webs of subtropical streams? Freshwater Biology 48: 1105–1116.

Clarke KR, Warwick RM. 2001. Change in marine communities: an approach to statistical analysis and interpretation. PRIMER-E Ltd, Plymouth, UK.

Coat S, Monti D, Bouchon C, Lepoint G. 2009. Trophic relationships in a tropical stream food web assessed by stable isotope analysis. Freshwater Biology 54: 1028–1041.

Cole JJ, Prairie YT, Caraco NF, McDowell WH, Tranvik LJ, Striegl RG, Duarte CM, Kortelainen P, Downing JA, Middelburg JJ, Melack J. 2007. Plumbing the Global Carbon Cycle: Integrating Inland Waters into the Terrestrial Carbon Budget. Ecosystems 10: 172-185.

Coley PD, ABarone J. 1996. Herbivory and plant defenses in tropical forests. Annual Review of Ecology and Systematics 27: 305 – 335.

Conway D, Persechino A, Ardoin-Bardin S, Hamandawana H, Dieulin C, Mahé G. 2009. Rainfall and Water Resources Variability in Sub-Saharan Africa during the Twentieth Century. Journal of Hydrometeorology 10: 41-59

Conway D. 2002. *Extreme rainfall events and Lake level changes in East Africa: Recent events and historical precedents*. The East African Great Lakes: Limnology, Palaeolimnology and Biodiversity, E. O. Odada and D. O. Olago, Eds , Advances in Global Change Research Series, Vol. 12, Kluwer, 63–92.

Corbet PS. 1961. The food of non-cichlid fishes in the Lake Victoria basin, with remarks on their evolution and adaptation to lacustrine conditions. Proceedings of the Zoological Society of London 136: 1–101.

Cory RM, McKnight DM. 2005. Fluorescence spectroscopy reveals ubiquitous presence of oxidized and reduced quinones in dissolved organic matter. Environmental Science & Technology 39: 8142-8149.

Cory RM, Miller MP. Mcknight DM, Guerard JJ, Miller PL. 2010. Effect of instrument-specific response on the analysis of fulvic acid fluorescence spectra. Limnology & Oceanography: Methods 8: 67–78, doi:10.4319/lom.2010.8.0067

COWI Consulting Engineers. 2002. Integrated Water Quality/ Limnological Study for Lake Victoria. Lake Victoria Environmental project, Part II Technical Report. Kisumu.

Cumberlidge N, Dobson M. 2008. A new species of freshwater crab of the genus *Potamonautes* Macleay, 1838 (Brachyura: Potamoidea: Potamonautidae) from the forested highlands of western Kenya, East Africa. Proceedings of the Biological Society of Washington 121: 468-474.

Cummins KW, Merritt RW, Andrade P. 2005. The use of invertebrate functional groups to characterize ecosystem attributes in selected streams and rivers in southeast Brazil. Studies on the Neotropical Fauna and Environment 40: 69–89.

Cummins KW, Petersen RC, Howard FO, Wuycheck JC, Holt VI. 1973. The utilization of leaf litter by stream detritivores. Ecology 54: 336–345.

Dangles OJ, Guérold FA. 2000. Structural and functional responses of benthic macroinvertebrates to acid precipitation in two forested headwater streams (Vosges Mountains, northeastern France). Hydrobiologia 418: 25–31.

Davis AM, Blanchette ML, Pusey BJ, Jardine TD, Pearson RG. 2012. Gut content and stable isotope analyses provide complementary understanding of ontogenetic dietary shifts and trophic relationships among fishes in a tropical river. Freshwater Biology 57: 2156-2172.

Day JA, de Moor IJ. 2002a. Guides to the freshwater invertebrates of southern Africa. Volume 5: Non-arthropods (the protozoans, Porifera, Cnidaria, Platyhelminthes, Nemertea, Rotifera, Nematoda, Nematomorpha, Gastrotrichia, Bryozoa, Tardigrada, Polychaeta, Oligochaeta and Hirudinea). WRC Report No. TT 167/02. Water Research Commission, Pretoria, South Africa.

Day JA, de Moor IJ. 2002b. Guides to the freshwater invertebrates of southern Africa. Volume 6: Arachnida and Mollusca (Araneae, Water Mites and Mollusca). WRC Report No. TT 182/02. Water Research Commission, Pretoria, South Africa.

Day JA, Harrison AD, de Moor IJ. 2002c. Guides to the Freshwater Invertebrates of Southern Africa, Volume 9: Diptera. WRC report No. TT 201/02, South Africa

DDP 2008b. District Development Plan 2002-2008, Bomet District, Kenya. Effective Management for Sustainable Economic Growth and Poverty Reduction. Government Printer, Nairobi.

DDP. 2008a. District Development Plan 2002-2008, Bomet District, Kenya. Effective Management for Sustainable Economic Growth and Poverty Reduction. Government Printer, Nairobi.

de Moor IJ, Day JA, de Moor FC. 2003a. Guides to the freshwater invertebrates of southern Africa. Volume 7: Insecta I: Ephemeroptera, Odonata and Plecoptera. WRC Report No. TT 207/03. Water Research Commission, Pretoria, South Africa.

de Moor IJ, Day JA, de Moor FC. 2003b. Guides to the freshwater invertebrates of southern Africa. Volume 8: Insecta II: Hemiptera, Megaloptera, Neuroptera, Trichoptera and Lepidoptera. WRC Report No. TT 214/03. Water Research Commission, Pretoria, South Africa.

de Oliveira ALH, Nessimian JL. 2010. Spatial distribution and functional feeding groups of aquatic insect communities in Serra da Bocaina streams, southeastern Brazil. Acta Limnologica Brasiliensia: 22: 424-441.

de Ruiter PC, Wolters V, Moore JC, Winemiller KO. 2005. Food web ecology: playing Jenga and beyond. Science 309: 68–71.

de Vos L. 2000. The non-cichlids of Lake Victoria system: diversity, taxonomy and identification problems. A Paper presented at Lake Victoria 2000: A New Beginning International Conference, Jinja, Uganda. 16th-19th May 2000.

Defersha M, Melesse AM. 2012. Field-scale investigation of the effect of land use on sediment yield and surface runoff using runoff plot data and models in the Mara River basin, Kenya, CATENA. Catena. doi:10.1016/j.catena.2011.07.010.

Delong MD, Thorp JH. 2006. Significance of instream autotrophs in trophic dynamics of the Upper Mississippi River. Oecologia 147: 76-85.

Dessu SB, Melesse AM, Bhat MG, McClain ME. 2014. Assessment of water resources availability and demand in the Mara River Basin. CATENA115: 104-114.

Dobson AP, Borner M, Sinclair AR, Hudson PJ, Anderson TM, Bigurube et al. 2010. Road will ruin Serengeti. Nature 467 (7313): 272-273.

Dobson M, Hildrew AG. 1992. A test of resource limitation among shredding detritivores in low order streams in southern England. Journal of Animal Ecology 61:69–77.

Dobson M, Magana AM, Lancaster J, Mathooko JM. 2007a. Aseasonality in the abundance and life history of an ecologically dominant freshwater crab in the Rift Valley, Kenya. Freshwater Biology 52: 215–225.

Dobson M, Magana AM, Mathooko JM, Ndegwa FK. 2007b. Distribution and abundance of freshwater crabs (Potamonautes spp.) in rivers draining Mt Kenya, East Africa. Fundamental and Applied Limnology 168: 271–279.

Dobson M, Magana MA, Mathooko JM, Ndegwa FK. 2002. Detritivores in Kenyan highland streams: more evidence for the paucity of shredders in the tropics? Freshwater Biology 47: 909–919.

Dobson M, Mathooko JM, Ndegwa FK, M'Erimba C. 2003. Leaf litter processing rates in a Kenyan highland stream, the Njoro River. Hydrobiologia 519: 207–210.

Dobson M. 2004. Freshwater crabs in Africa. Freshwater Forum 21: 3–26.

Donohue I, Irvine K. 2004. Seasonal patterns of sediment loading and benthic invertebrate community dynamics in Lake Tanganyika, Africa. Freshwater Biology 49: 320–331.

Douglas MM, Bunn SE, Davies PM. 2005. River and wetland food webs in Australia's wet-dry tropics: general principles and implications for management. Marine and Freshwater Research 56: 329–342.

du Toit JT, Bryant JP, Frisby K. 1990. Regrowth and palatability of Acacia shoots following pruning by African savanna browsers. Ecology 71: 149–54.

du Toit JT, Cumming DHM. 1999. Functional significance of ungulate diversity in African savannas and the ecological implications of the spread of pastoralism. Biodiversity and Conservation 8: 1643–1661.

Dudgeon D, Arthington AH, Gessner MO, Kawabata Z, Knowler DJ, Lêvêque C, Naiman RJ, Prieur-Richard A-H, Soto D, Stiassny MLJ, Sullivan CA. 2006. Freshwater biodiversity: importance, threats, status and conservation challenges. Biological Reviews 81: 163–182.

Dudgeon D, Cheung FKW, Mantel SK. 2010. Foodweb structure in small streams: do we need different models for the tropics? Journal of the North American Benthological Society 29: 395-412.

Dudgeon D. 1999. Tropical Asian Streams: Zoobenthos, Ecology and Conservation. Hong Kong University Press, Hong Kong.

Dudgeon D. 2008. Tropical stream ecology (ed), Academic Press, London

Dudgeon D. 2010. Prospects for sustaining freshwater biodiversity in the 21st century: linking ecosystem structure and function. Current Opinion in Environmental Sustainability 2: 422–430.

Dudgeon D. 1999. Tropical Asian Streams: Zoobenthos, Ecology and Conservation, Hong Kong University Press, Hong Kong.

Dutton CL. 2012. Sediment fingerprinting in the Mara River: uncovering relationships between wildlife, tourism, and non-point source pollution. MSc. thesis, Yale University.

EAFRO (East African Fisheries Research Organization). 1955. Annual Report 1954/1955. Argus Ltd. Kampala.

Elmore HL, West WF. 1961. Effect of water temperature on stream reaeration. Journal of the Sanitary Engineering Division. Proceedings of the American Society of Civil Engineering, 87 (SA6): 59-71.

Elosegi A, Johnson LB. 2003. Wood in streams and rivers in developed landscapes. In: Gregory SV, Boyer, K.L., Gurnell, A.M. (eds) The ecology and management of wood in world rivers, 337-354 pp. American Fisheries Society, Bethesda, Maryland.

Elsdon TS, Wells BK, Campana SE, Gillanders BM, Jones CM, Limburg KE, Sector DH, Thorrold SR, Walther BD. 2008. Otolith chemistry to describe movements and life-history parameters of fishes:

hypotheses, assumptions, limitations and inferences. Oceanography and marine biology: an annual review 46: 297-330.

Ewing SA, Sanderman J, Baisden W, Wang Y, Amundson R. 2006. Role of large-scale soil structure in organic carbon turnover: evidence from California grassland soils. Journal of Geophysical Research 111:G03012.

FAO. 2010. Food and Agriculture Organization of the United Nations: Global Forest Resources Assessment Main report. FAO Forestry Paper 163, Food and Agriculture Organization of the United Nations: Rome.

Farjalla VF, Marinho CC, Faria BM, Amado AM, Esteves FDA, Bozelli RL, Giroldo D. 2009. Synergy of fresh and accumulated organic matter to bacterial growth. Microbial Ecology 57: 657-666.

Fellman JB, D'Amore DV, Hood E. 2008. An evaluation of freezing as a preservation technique for analyzing dissolved organic C, N and P in surface water samples. Science of the Total Environment 392: 305–12.

Fellman JB, Hood E, Spencer RGM. 2010. Fluorescence spectroscopy opens new windows into dissolved organic matter dynamics in freshwater ecosystems: A review. Limnology & Oceanography 55: 2452-2462.

Fellows CS, Clapcott JE, Udy JW, Bunn SE, Harch BD, Smith MJ, Davies PM. 2006. Benthic metabolism as an indicator of stream ecosystem health. Hydrobiologia 572: 71-87.

Ferreira V, Elosegi A, Gulis V, Pozo J, Graça MAS. 2006. Eucalyptus plantations affect fungal communities associated with leaf-litter decomposition in Iberian streams. Archiv fur Hydrobiologie. 166: 467–490.

Ferreira, V., A. C. Encalada, and M. A. S. Graça. 2012. Effects of litter diversity on decomposition and biological colonization of submerged litter in temperate and tropical streams. Freshwater Science: 31: 945–962.

Findlay S. 2003. Bacterial responses to variation in dissolved organic matter. In: Aquatic ecosystems: interactivity of dissolved organic matter, Findlay SEG, Sinsabaugh RL, (eds). Academic Press: New York; 363–382.

Finlay JC, Kendall C. 2007. Stable isotope tracing of temporal and spatial variability in organic matter sources to freshwater ecosystems. In: Michener RH, Lajtha K. (Eds), Stable isotopes in ecology and environmental science, 283–333 pp. 2nd edn. Malden: Blackwell.

Finlay JC. 2004. Patterns and controls of lotic algal stable carbon isotope ratios. Limnology & Oceanography 49: 850–861.

Fisher SJ, Brown ML, Willis DW. 2001. Temporal food web variability in an upper Missouri River backwater: energy origination points and transfer mechanisms. Ecology of Freshwater Fish 10: 154-167.

Flecker AS, McIntyre PB, Moore JW, Anderson JT, Taylor BW, Hall RO. 2010. Migratory fishes as material and process subsidies in riverine ecosystems.

Foley JA, DeFries R, Asner GP, Barford C, Bonan G, Carpenter SR. et al. 2005. Global consequences of land use. Science 309: 570-574.

France RL. 1997. Stable carbon and nitrogen isotopic evidence for ecotonal coupling between boreal forests and fishes. Ecology of Freshwater Fish 6: 78-83.

Frissell CA, Liss WJ, Warren CE, Hurley MD. 1986. A hierarchical framework for stream habitat classification: viewing streams in a watershed context. Environmental Management 10: 199-214.

Fry B, Sherr EB. 1989. $\delta^{13}C$ Measurements as indicators of carbon flow in marine and freshwater ecosystems. In Rundel PW, Ehleringer JR & Nagy KA (eds), Stable Isotopes in Ecological Research. Ecological Studies. springer-Verlag, New York, 196-229 pp.

Galois R, Richard P, Fricourt B. 1996. Seasonal variations in suspended particulate matter in the Marennes-Oleron Bay, France, using lipids as biomarkers. Estuaries, Coasts & Shelf Science 43: 335-357.

Gawne B, Merrick C, Williams DG, Rees G, Oliver R, Bowen PM, Treadwell S, Beattie G, Ellis I, Frankenberg J, Lorenz Z. 2007. Patterns of primary and heterotrophic productivity in an arid lowland river. River Research and Applications 23: 1070–1087.

Genereux DP, Hemond HF. 1992. Determination of gas exchange rate constants for a small stream on Walker Branch Watershed, Tennessee. Water Resources Research 28: 2365-2374.

Gereta E, Wolanski E, Borner M, Serneels S. 2002. Use of an ecohydrological model to predict the impact on the Serengeti ecosystem of deforestation, irrigation and the proposed Amala weir water diversion project in Kenya. Ecohydrology and Hydrobiology 2: 127-134.

Gereta E, Wolanski E. 1998. Wildlife-water quality interactions in the Serengeti National Park, Tanzania. African Journal of Ecology 36: 1–14.

Gessner, M.O., Chauvet, E., 2002. A case for using litter breakdown to assess functional stream integrity. Ecological Applications 12: 498–510.

Getenga ZM, Keng'ara FO, Wandiga SO. 2004. Determination of organochlorine pesticide residues in soil and water from River Nyando drainage system within Lake Victoria basin, Kenya. Bulletin of Environmental Contaminanation and Toxicology 72: 335-343.

Gichuki J, Guebas FD, MugoJ, Rabuor CO, Triest L, Dehairs F. 2001a. Species inventory and the local uses of the plants and fishes of the Lower Sondu Miriu wetland of Lake Victoria, Kenya. Hydrobiologia 458: 99–106.

Gichuki JW., Triest L, Dehairs F. 2001b. The use of stable carbon isotopes as tracers of ecosystem functioning in contrasting wetland ecosystems of Lake Victoria, Kenya. Hydrobiologia 458: 91–97.

Gonçalves JF, Graça MAF, Callisto M. 2006. Leaf-litter breakdown in 3 streams in temperate, Mediterranean, and tropical Cerrado climates. Journal of the North American Benthological Society 24: 344–355.

Gophen M, Ochumba PBO, Kaufman LS. 1995. Some aspects of perturbation in the structure and biodiversity of the ecosystem of Lake Victoria (East Africa). Aquatic Living Resources 8: 27–41.

Goudswaard PC, Witte F, Katunzi EFB. 2008. The invasion of an introduced predator, Nile perch (*Lates niloticus*, L.), in Lake Victoria (East Africa): chronology and causes. Environmental Biology of Fishes 81: 127–139.

Goudswaard PC, Witte F, Wanink JH. 2006. The shrimp Caridina nilotica in Lake Victoria (East Africa), before and after the Nile perch increase. Hydrobiologia 563: 31–44.

Goudswaard PC, Witte F. 1997. The catfish fauna of Lake Victoria after the Nile perch upsurge. Environmental Biology of Fishes 49: 21–43.

Government of the Republic of Kenya (GOK). 2007. Kenya Vision 2030, Government printer, Nairobi.

Government of the Republic of Kenya (GOK). 2008. First medium term plan, 2008 – 2012, Kenya vision 2030, Globally Competitive and Prosperous Kenya, Government printer, Nairobi.

Government of the Republic of Kenya (GOK). 2009. Rehabilitation of the Mau Forest Ecosystem, A Project concept prepared by the interim coordinating secretariat, Office of the Prime Minister, on behalf of the Government of Kenya, Government printer, Nairobi.

Graça MAS, Cressa C, Gessner MO, Feio MJ, Callies KA, Barrios C. 2001. Food quality, feeding preferences, survival and growth of shredders from temperate and tropical streams. Freshwater Biology 46: 1–11.

Graça MAS, Pozo J, Canhoto C, Elosegui A. 2002. Effects of Eucalyptus plantations on detritus, decomposers and detritivores in streams. Scientific World 2: 1173–85.

Graeber D, Gelbrecht J, Pusch MT, Anlanger C, von Schiller D. 2012. Agriculture has changed the amount and composition of dissolved organic matter in Central European headwater streams. Science of the Total Environment 438: 435-446.

Graham M. 1929. The Victoria Nyanza and its Fisheries – A Report on the Fishing Surveys of Lake Victoria (1927–1928). Crown Agents Colonies: London.

Greathouse EA, Pringle CM. 2006. Does the river continuum concept apply on a tropical island? Longitudinal variation in a Puerto Rican stream. Canadian Journal of Fisheries and Aquatic Science 63: 134–152.

Gregory KJ, Walling DE. 1973. Drainage Basin Form and Process: Geomorphological Approach. Edward Arnold, London.

Grey J, Harper DM. 2002. Using stable isotope analyses to identify allochthonous inputs to Lake Naivasha mediated via the hippopotamus gut. Isotopes Environ. Health Studies 38: 245–250.

Griffiths NA, Tank JL, Royer TV, Roley SS, Rosi-Marshall EJ, Whiles MR, Beaulieu JJ, Johnson LT. 2013. Agricultural land use alters the seasonality and magnitude of stream metabolism. Limnology & Oceanography 58: 1513-1529.

Gücker B, Boëchat IG, Giani A. 2009. Impacts of agricultural land use on ecosystem structure and whole-stream metabolism of tropical Cerrado streams. Freshwater Biology 54: 2069–2085.

Guenet B, Danger M, Abbadie L, Lacroix G. 2010. Priming effect: bridging the gap between terrestrial and aquatic ecology. Ecology 91: 2850-2861.

Gulis V, Farreira V, Graça MAS. 2006. Stimulation of leaf litter decomposition and associated fungi and invertebrates by moderate eutrophication:implications for stream assessment. Freshwater Biology 51: 1655–1669.

Gulis V, Suberkropp K. 2003. Leaf litter decomposition and microbial activity in nutrient-enriched and unaltered reaches of a headwater stream. Freshwater Biology 48: 123-134.

Hadwen WL, Fellows CS, Westhorpe DP, Rees GN, Mitrovic SM, Taylor B, Baldwin DS, Silvester E, Croome R. 2010a. Longitudinal trends in river functioning: patterns of nutrient and carbon processing in three Australian rivers. River Research and Applications 26: 1129–1152.

Hadwen WL, Spears M, Kennard MJ. 2010b. Temporal variability of benthic algal $\delta^{13}C$ signatures influences assessments of carbon flows in stream food webs. Hydrobiologia 651: 239–251.

Halbedel S, Büttner O, Weitere M. 2013. Linkage between the temporal and spatial variability of dissolved organic matter and whole-stream metabolism. Biogeosciences 10: 5555–5569.

Hall RO, Likens GE, Malcom HM et al 2001. Trophic basis of invertebrate production in 2 streams at the Hubbard Brook Experimental Forest. Journal of the North American Benthololological Society 20: 432–447.

Hall RO, Tank JL. 2003. Ecosystem metabolism controls nitrogen uptake in streams in Grand Teton National Park, Wyoming. Limnology and Oceanography 48: 1120-1128.

Hallam, A., and J. Read. 2006. Do tropical plant species invest more in anti-herbivore defence than temperate species? A test in *Eucryphia* (Cunoniaceae) in eastern Australia. Journal of Tropical Ecology 22: 41–51.

Hammer Ø, Harper DAT, Ryan PD. 2001. PAST: Paleontological Statistics Software Package for Education and Data Analysis. Palaeontologia Electronica 4(1).

Harper DM, Adams C, Mavuti K. 1995. The aquatic plant communities at the Lake Naivasha wetland, Kenya: pattern, dynamics and conservation. Wetlands Ecology and Management 3: 111–123.

Harrington RR, Kennedy BP, Chamberlain CP, Blum JD, Folt CL. 1998. [15]N enrichment in agricultural catchments: field patterns and applications to tracking Atlantic salmon (*Salmo salar*). Chemical Geology 147: 281–294.

Harrison AD, Hynes HBN. 1988. Benthic fauna of Ethiopian streams and rivers. Arch. Hydrobiol. 81: 1-36.

Hecky RE, Bugenyi FWB, Ochumba POB., Talling JF, Mugidde R, Gophen M, Kaufman L. 1994. Deoxygenation of the hypolimnion of Lake Victoria. Limnology and Oceanography 39: 1476–1481.

Hecky RE, Muggide R, Ramlal PS, Talbot MR, Kling GW. 2010. Multiple stressors cause rapid ecosystem change in Lake Victoria. Freshwater Biology 55: 19-42.

Hecky RE. 1993. The eutrophication of Lake Victoria. Verhandlungen des Internationalen Verein Limnologie 25: 39-48.

Heeg J, Breen CM. 1982. Man and the Pongolo Floodplain. In: SANS Programmes, editor. Council for Scientific and Industrial Research, Pretoria.

Heino J. 2009. Biodiversity of aquatic insects: spatial gradients and environmental correlates of assemblage-level measures at large scales. Freshwater Reviews 2: 1–29.

Helms JR, Stubbins A, Ritchie JD, Minor EC, Kieber DJ, Mopper K. 2008. Absorption spectral slopes and slope ratios as indicators of molecular weight, source, and photobleaching of chromophoric dissolved organic matter. Limnology & Oceanography 53: 955-969.

Herwig BR, Wahl DH, Dettmers JM. Soluk DA. 2007. Spatial and temporal patterns in the food web structure of a large floodplain river assessed using stable isotopes. Canadian Journal of Fisheries and Aquatic Sciences 64: 495-508.

Hieber M, Gessner MO. 2002. Contribution of stream detrivores, fungi, and bacteria to leaf breakdown based on biomass estimates. Ecology 83: 1026–1038.

Hladyz S, Åbjörnsson K, Cariss H, Chauvet E, Dobson M, Elosegi A, Ferreira V, Fleituch T, Gessner MO, Giller PS, Gulis V, Hutton SA, Lacoursiere JO, Lamothe S, Lecerf A, Malmqvist B, Mckie BG, Nistorescu M, Preda E, Riipnen MP, Rîşnoveanu G, Schindler M, Tiegs SD, Vought LB-M, Woodward G. 2011. Stream ecosystem functioning in an agriculture landscape: the importance of terrestrial-aquatic linkages. Adv. Ecol. Res. 44: 211–276.

Hladyz S, Tiegs SD, Gessner MO, Giller PS, Rîşnoveanu G, Preda E, Nistorescu M, Schindler M, Woodward G. 2010. Leaf-litter breakdown in pasture and deciduous woodland streams: A comparison among three European regions. Freshwater Biology 55:1916–1929.

Hoberg P, Lindholm M, Ramberg L, Hessen DO. 2002. Aquatic food-web dynamics on a floodplain in the Okavango Delta, Botswana. Hydrobiologia 470: 23–30.

Hoeinghaus DJ, Winemiller KO, Agostinho AA. 2007. Landscape-scale hydrologic characteristics differentiate patterns of carbon flow in large-river food webs. Ecosystems 10: 1019:1033.

Hoffman CM, Melesse AM, McClain ME. 2011. Geospatial mapping and analysis of water availability-demand-use within the Mara River Basin. In: Melesse AM. (Ed), Nile River Basin: Hydrology, Climate and Water Use. Springer, Dordrecht. New York, pp. 359–382.

Hudson N, Baker A, Reynolds DM, Carliell-Marquet C, Ward D. 2009. Changes in freshwater organic matter fluorescence intensity with freezing/ thawing and dehydration/ rehydration. Journal of Geophysical Research 114: G00F08.

Huguet A, Vacher L, Relexans S, Saubusse S, Parlanti E, Froidefond JM. 2009. Properties of fluorescent dissolved organic matter in the Gironde Estuary. Organic Geochemistry 40: 706-719.

Hunt RJ, Jardine TD, Hamilton SK, Bunn SE. 2012. Temporal and spatial variation in ecosystem metabolism and food web carbon transfer in a wet-dry tropical river. Freshwater Biology 57: 435-450.

Huryn AD, Riley RH, Young RG, Arbuckle CJ, Peacock K, Lyons G. 2001. Temporal shift in the contribution of terrestrial organic matter to consumer production in a grassland river. Freshwater Biology 46: 213–226.

Hynes HBN. 1975. The stream and its Valley. Verhandlungen der Internationale Vereinigung für Theoretische und Angewandte Limnologie 19: 1-15.

Irons JG, Oswood MW, Stout RJ, Pringle CM. 1994. Latitudinal patterns in leaf litter breakdown: is temperature really important? Freshwater Biology 32: 401–411.

Jackson AJH, McCarter PS. 1994. A profile of the Mau complex. KIFCON, Nairobi

Jackson MC, Donohue I, Jackson AL, Britton JR, Harper DM, Grey J. 2012. Population-level metrics of trophic structure based on stable isotopes and their application to invasion ecology. PloS one 7: e31757.

Jacobs SM, Bechtold JS, Biggs HC, Grimm NB, Lorentz S, McClain ME, Naiman RJ, Perakis SS, Pinay G, Scholes MC. 2007. Nutrient vectors and riparian processing: A review with special reference to African semiarid Savanna ecosystems. Ecosystems 10: 1231-1249.

Jacobsen D, Cressa C, Dudgeon D. 2008. Macroinvertebrates: composition, life histories and production. *In* D. Dudgeon (editor). Aquatic ecosystems: tropical stream ecology, 65–105 pp. Elsevier Science, London, UK.

Jaetzold R, Schmidt H. 1983. Farm Management Handbook of Kenya: natural conditions and farm management information - Central Kenya, Vol II/B. Ministry of Agriculture, Kenya, in Cooperation with the Germany Agricultural Team (GAT) of the Germany Agency for Technical Cooperation (GTZ).

Jaffé R, McKnight D, Maie N, Cory R, McDowell WH, Campbell JL. 2008. Spatial and temporal variations in DOM composition in ecosystems: The importance of long-term monitoring of optical properties. Journal of Geophysical Research 113.

Jardine T, Pusey B, Hamilton S, Pettit N, Davies P, Douglas M, Sinnamon V, Halliday I, Bunn S. 2011. Fish mediate high food web connectivity in the lower reaches of a tropical floodplain river. Oecologia 168: 829-838.

Jardine TD, Pettit NE, Warfe DM, Pusey BJ, Ward DP, Douglas MM, Davies PM, Bunn SE. 2012. Consumer–resource coupling in wet–dry tropical rivers. Journal of Animal Ecology 81: 310-322.

Jepsen DB, Winemiller KO. 2007. Basin geochemistry and isotopic ratios of fishes and basal production sources in four neotropical rivers. Ecology of Freshwater Fish 16: 267-281.

JICA (Japan International Cooperation Agency) 1992. The Study of the National Water Master Plan. Japan International Cooperation Agency. Ministry of Water Resources and management, Nairobi.

Jinggut T, Yule CM, Boyero L. 2012. Stream ecosystem integrity is impaired by logging and shifting agriculture in a global megadiversity center (Sarawak, Borneo). Science of the Total Environment 437: 83–90.

Julian JP, Doyle MW, Stanley EH. 2008. Empirical modeling of light availability in rivers. Journal of Geophysical Research-Biogeosciences 113: G03022.

Julian JP, Seegert SZ, Powers SM, Stanley EH, Doyle MW. 2011. Light as a first-order control on ecosystem structure in a temperate stream. Ecohydrology 4: 422–432.

Junk WJ, Bayley PB, Sparks RE. 1989. The flood pulse concept in river-floodplain systems. In Proceedings of the International Large River Symposium, Vol. 106, Dodge DP (ed.). Canadian Special Publication in Fisheries and Aquatic Sciences 106: 110–127.

Kalbitz K, Schwesig D, Schmerwitz J, Kaiser K, Haumaier L, Glaser B, Ellerbrock R, Leinweber P. 2003. Changes in properties of soil-derived dissolved organic matter induced by biodegradation. Soil Biolology & Biochemistry 35: 1129–1142.

Kanga EM, Ogutu J, Hans-Peter P, Olff H. 2012. Human-hippo conflicts in Kenya during 1997-2008: Vulnerability of a megaherbivore to anthropogenic land use changes. Journal of Land Use Science 7: 395-406.

Kanga EM, Ogutu JO, Olff H, Santema P. 2011. Population trend and distribution of the vulnerable common hippopotamus *Hippopotamus amphibius* in the Mara Region of Kenya. Oryx 45: 20-27.

Kasangaki A, Chapman LJ, Balirwa J. 2008. Land use and the ecology of benthic macroinvertebrate assemblages of high-altitude rainforest streams in Uganda. Freshwater Biology 53: 681–697.

Kashian, D.R., Zuellig, R.E., Mitchell, K.A., Clements, W.H., 2007. The cost of tolerance: sensitivity of stream benthic communities to UV-B and metals. Ecological Applications 17: 365–75.

Kaushal SS, Delaney-Newcomb K, Findlay SE, Newcomer TA, Duan S, Pennino MJ, Belt KT. 2014. Longitudinal patterns in carbon and nitrogen fluxes and stream metabolism along an urban watershed continuum. Biogeochemistry 1-22.

Kennard MJ, Pusey BJ, Olden JD, Mackay SJ, Stein JL, Marsh N. 2010. Classification of natural flow regimes in Australia to support environmental flow management. Frashwater Biology 55:171-193.

KFS (Kenya Forestry Service) 2009. A Guide to On-Farm Eucalyptus Growing in Kenya.

Kiambi S, Kuloba B, Kenana L, Muteti D, Mwenda E. 2012. Wet Season Aerial Count of Large Herbivores in Masai Mara National Reserve and the Adjacent Community Areas (June 2010). Mara Research Station, Kenya Wildlife Service, Narok, Kenya.

Kibichii S, Shivoga WA, Muchiri M, Miller SN. 2007. Macroinvertebrate assemblages along a land-use gradient in the upper River Njoro watershed of Lake Nakuru drainage basin, Kenya. Lakes & Reservoirs: Research and Management 12: 107–117.

Kilonzo F, Masese FO, Van Griensven A, Bauwens W, Obando J, Lens PNL. 2013. Spatial–temporal variability in water quality and macro-invertebrate assemblages in the Upper Mara River basin, Kenya. Phys. Chem. Earth PT A/B/C 67-69: 93-104.

Kilonzo FN. 2014. Assessing the Impacts of Environmental Changes on the Water Resources of the Upper Mara, Lake Victoria Basin. PhD thesis, Vrije Universiteit Brussel and UNESCO-IHE Institute for Water Education, Delft.

Kitaka N, Harper DM, Mavuti KM, Pacini N. 2002. Chemical characteristics, with particular reference to phosphorus, of the rivers draining into Lake Naivasha, Kenya. In Lake Naivasha, Kenya (pp. 57-71). Springer Netherlands.

Kitchell JF, Oneill RV, Webb D, Gallepp GW, Bartell SM, Koonce JF, Ausmus BS. 1979. Consumer regulation of nutrient cycling. BioScience 29: 28-34.

Kizza M, Rodhe A, Xu C, Ntale HK, Halldin S. 2009. Temporal rainfall variability in the Lake Victoria Basin in East Africa during the twentieth century. Theoretical and Applied Climatology 98: 119–135.

Kline TC, Goering JJ, Mathisen OA, Poe PH, Parker PL. 1990. Recycling of elements transported upstream by runs of Pacific salmon: I. δ^{15} N and δ^{13} C evidence in Sashin Creek, southeastern Alaska. Canadian Journal of Fisheries and Aquatic Sciences 47: 136–144.

KNBS-IHBS. 2007. Kenya National Bureau of Statistics (KNBS)/ Kenya Integrated Household Budget Survey (KIHBS)—2005/06. Ministry of Planning and National Development, Nairobi.

KNBS-LS, 2009. Livestock Population by Type and District. Census KNBS, Nariobi.

Kolding J, van Zwieten P, Mkumbo O, Silsbe G, Hecky RE. 2008. Are the Lake Victoria fisheries threatened by exploitation or eutrophication? Towards an ecosystem based approach to management. In: The Ecosystem Approach to Fisheries (Eds G. Bianchi & H.R. Skjodal), pp. 309–354. CAB International, Rome.

Lambert T, Pierson-Wickmann AC, Gruau G, Thibault JN, Jaffrezic A. 2011. Carbon isotopes as tracers of dissolved organic carbon sources and water pathways in headwater catchments. Journal of Hydrology 402: 228–238.

Lamberti GA, Chaloner DT, Herschy AE. 2010. Linkages among aquatic ecosystems. Journal of the North American Benthological Society 29(1): 245–263.

Lambin EF, Geist HJ. 2006. Land –use and land-cover change: local processes and global impacts, Springer, 222 pp.

Lamprey RH, Reid RS. 2004. Expansion of human settlement in Kenya's Maasai Mara: what future for pastoralism and wildlife? Journal of Biogeography 31: 997-1032.

Lancaster J, Dobson M, Magana AM, Arnold A, Mathooko JM. 2008. An unusual trophic subsidy and species dominance in a tropical stream. Ecology 89: 2325-2334.

Lau DCP, Leung KMY, Dudgeon D. 2009. What does stable isotope analysis reveal about food webs and trophic relationships in tropical streams? A synthetic study from Hong Kong. Freshwater Biology 54: 127–141.

Layman CA, Arrington DA, Montaná CG, Post DM. 2007. Can stable isotope ratios provide quantitative measures of trophic diversity within food webs? Ecology 88: 42–48.

Lecerf A, Usseglio-Polatera P, Charcosset JY, Lambrigot D, Bracht B, Chauvet E. 2006. Assessment of functional integrity of eutrophic streams using litter breakdown and benthic macroinvertebrates. Archiv für Hydrobiologie 165: 105–126.

Lefrançois E, Coat S, Lepoint G, Vachiéy N, Gros O, Monti D. 2011. Epilithic biofilm as a key factor for small-scale river fisheries on Caribbean islands. Fisheries Management and Ecology 18: 211-220.

Lehman, J. T. and D. K. Branstrator, 1994: Nutrient dynamics and turnover rates of phosphate and sulfate in Lake Victoria, East Africa. Limnology & Oceanography 39: 227–233.

Leroux SJ, Loreau M. 2008. Subsidy hypothesis and strength of trophic cascades across ecosystems. Ecology Letters 11: 1147-1156.

Leung ASL, Li AOY, Dudgeon D. 2012. Scales of spatiotemporal variation in macroinvertebrate assemblage structure in monsoonal streams: the importance of season. Freshwater Biology 57: 218–231.

Lévêque C. 1995. Role and consequences of fish diversity in the functioning of African freshwater ecosystems: a review. Aquatic Living Resources 8: 59-78.

Lewis W. 2008. Physical and chemical features of tropical flowing waters. In: Dudgeon D (ed), Tropical Stream Ecology, 1-22 pp. Academic Press: San Diego

Lewis WMJ, Hamilton SK, Rodriguez MA, Saunders JFI, Lasi MA. 2001. Foodweb analysis of the Orinoco floodplain based on production estimates and stable isotope data. Journal of the North American Benthological Society 20: 241–254.

Li AOY, Dudgeon D. 2008. Food resources of shredders and other benthic macroinvertebrates across a range of shading conditions in tropical Hong Kong streams. Freshwater Biology 53: 2011–2025.

Ligeiro R, Moretti MS, Gonçalves JFJr, Callisto M. 2010. Whatis more important for invertebrate colonization in a stream with low-quality litter inputs: exposure time or leaf species? Hydrobiologia 654: 125–136.

176

Logan JM, Jardine TD, Miller TJ, Bunn SE, Cunjak RA, and Lutcavage ME. 2008. Lipid corrections in carbon and nitrogen stable isotope analyses: comparison of chemical extraction and modelling methods. Journal of Animal Ecology 77: 838–846.

Lovett JC, Wasser SK (eds). 1993. Biogeography and Ecology of the Rain Forests of Eastern Africa. Cambridge University Press, Cambridge.

Lowe-McConnell RH. 1987. Ecological Studies in Tropical Fish Communities. University Press: Cambridge, UK.

Ludwig W, Suche, PA, Probst JL. 1996. River discharges of carbon to the world's oceans: Determining local inputs of alkalinity and of dissolved and particulate organic carbon, CR Acad. Sci. II A, 323: 1007–1014.

Lung'ayia H, Sitoki L, Kenyanya M. 2001. The nutrient enrichment of Lake Victoria (Kenyan waters). Hydrobiologia 458: 75–82.

LVBC and WWF-ESARPO (Lake Victoria Basin Commission and WWF-Eastern and Southern Africa Regional Programme Office) (2010). Assessing Reserve Flows for the Mara River. Nairobi and Kisumu, Kenya.

M'Erimba CM, Mathooko JM, Leichtfried M. 2006. Variations in coarse particulate organic matter in relation to anthropogenic trampling on the banks of the Njoro River, Kenya. African Journal Ecology 44: 282–285.

Machiwa JF. 2010. Stable carbon and nitrogen isotopic signatures of organic matter sources in near-shore areas of Lake Victoria, East Africa. Journal of Great Lakes Research 36: 1–8.

Magana AEM, Bretshko G. 2003. Retention of Coarse Particulate Organic Matter on the Sediments of Njoro River, Kenya. International Review of Hydrobiolgy 88: 414–426.

Magana AEM. 2001. Litter input from riparian vegetation to streams: a case study of the Njoro River, Kenya. Hydrobiologia 458:141–149.

Magana AM, Dobson M, Mathooko JM. 2012. Modifying Surber sampling technique increases capture of freshwater crabs in African upland streams. Inland Waters 2: 11-15.

Maitama JM, Mugatha SM, Reid RS, Gachimbi LN, Majule A, Lyaruu H, Pomery D, Mathai S, Mugisha S. 2009. The linkages between land use change, land degradation and biodiversity across East Africa. African Journal of Environmental Science and Technology 3: 310-325.

MALP (2009a). Ministry of Agriculture and Livestock Production, Narok South District, Kenya.

MALP (2009b). Ministry of Agriculture and Livestock Production, Bomet District, Kenya.

Mango LM, Melesse AM, McClain ME, Gann D, Setegn SG. 2011. Land use and climate change impacts on the hydrology of the upper Mara River Basin, Kenya: Results of a modeling study to support better resource management. Hydrology and Earth System Sciences 15: 2245–2258.

Mann HB, Whitney DR. 1947. On a test of whether one of two random variables is stochastically larger than the other. The annals of mathematical statistics 50-60.

Mantel SK, Salas M, Dudgeon D. 2004. Food web structure in a tropical Asian forest stream. Journal of the North American Benthological Society 23: 728–755.

Manyala JO, Bolo JZ, Onyang'o S and Rambiri PO. 2005. *Indigenous kwnoledge and baseline data survey on fish breeding areas and seasons in Lake Victoria, Kenya.* In: Knowledge and Experiences gained from Managing the Lake Victoria Ecosystem, Mallya GA, Katagira FF, Kang'oha G, Mbwana SB, Katunzi EF,Wambede JT, Azza N,Wakwabi E, Njoka SW, Kusewa M, Busulwa H (eds). Regional Secretariat, Lake Victoria Environmental Management Project (LVEMP): Dar es Salaam; 529-551.

Marcarelli AM, Baxter CV, Mineau MM, Hall RO. 2011. Quantity and quality: unifying food web and ecosystem perspectives on the role of resource subsidies in freshwaters. Ecology 92: 1215-1225.

March JG, Pringle CM. 2003. Food web structure and basal resource utilization along a tropical island stream continuum, Puerto Rico. Biotropica 35: 84–93.

Marwick TR, Borges AV, Acker KV, Darchambeau F, Bouillon S. 2014b. Disproportionate Contribution of Riparian Inputs to Organic Carbon Pools in Freshwater Systems. Ecosystems 17: 974-989.

Marwick TR, Tamooh F, Ogwoka B, Teodoru C, Borges AV, Darchambeau F, Bouillon S. 2014a. Dynamic seasonal nitrogen cycling in response to anthropogenic N-loading in a tropical catchment, Athi–Galana–Sabaki River, Kenya. Biogeosciences 11: 443–460.

Marzolf ER, Mulholland PJ, Steinman AD. 1994. Improvements to the diurnal upstream-downstream dissolved oxygen change technique for determining whole-stream metabolism in small streams. Canandian Journal of Fisheries and Aquatic Sciences 51.

Masese FO, Kitaka N, Kipkemboi J, Gettel GM, Irvine K, McClain ME. 2014a. Macroinvertebrate functional feeding groups in Kenyan highland streams: evidence for a diverse shredder guild. Freshwater Science 33: 435-450.

Masese FO, Kitaka N, Kipkemboi J, Gettel GM, Irvine K, McClain ME. 2014b. Litter processing and shredder distribution as indicators of riparian and catchment influences on ecological health of tropical streams. Ecological Indicators 46: 23–37.

Masese FO, Abrantes KG, Gettel GM, Bouillon S, Irvine K, McClain ME. 2015. Are large herbivores vectors of terrestrial subsidies for riverine food webs? *Ecosystems,* in press.

Masese FO, McClain ME. 2012. Trophic resources and emergent food web attributes in rivers of the Lake Victoria Basin: a review with reference to anthropogenic influences. Ecohydrology 5: 685–707.

Masese FO, Muchiri M, Raburu PO. 2009a. Macroinvertebrate assemblages as biological indicators of water quality in the Moiben River, Kenya. African Journal of Aquatic Science 34: 15–26.

Masese FO, Raburu PO, Muchiri M 2009b. A preliminary benthic macroinvertebrate index of biotic integrity (B-IBI) for monitoring the Moiben River, Lake Victoria Basin, Kenya. African Journal of Aquatic Science 34: 1–14.

Masese FO, Raburu PO, Mwasi BN and Etiégni L. 2011. Effects of Deforestation on Water Resources: Integrating Science and Community Perspectives in the Sondu-Miriu River Basin, Kenya. In: *New Advances and Contributions to Forestry Research*, Oteng-Amoako AA (ed.), 1-18 pp. InTech Publisher, Riejka.

Mathooko JM, Kariuki ST. 2000. Disturbances and species distribution of the riparian vegetation of a Rift Valley stream. African Journal of Ecology 38: 123-129.

Mathooko JM, Magana AM, Nyang'au IM. 2000. Decomposition of Syzygium cordatum leaves in a Rift Valley stream ecosystem. African Journal of Ecology 38: 265–368.

Mathooko JM, Mavuti KM. 1992. Composition and seasonality of benthic invertebrates, and drift in the Naro Moru River, Kenya. Hydrobiologia 232: 47–56.

Mathooko JM, M'Erimba CM, Kipkemboi J, Dobson M. 2009. Conservation of highland streams in Kenya: the importance of the socio-economic dimension in effective management of resources. Freshwater Reviews 2: 153-165.

Mathooko JM, Morara GO, Leichtfried M. 2001. Leaf litter transport and retention in a tropical Rift Valley stream: an experimental approach. Hydrobiologia 443: 9–18.

Mathooko JM, Mpawenayo B, Kipkemboi JK, M'Erimba CM. 2005. Distributional Patterns of Diatoms and *Limnodrilus* Oligochaetes in a Kenyan Dry Streambed Following the 1999–2000 Drought Conditions. International Review of Hydrobiology 90: 185–200.

Mathooko JM. 2001. Disturbance of a Kenyan Rift Valley stream by the daily activities of local people and their livestock. Hydrobiologia 458: 131–139.

Mathuriau C, Chauvet E. 2002. Breakdown of leaf litter in a neotropical stream. Journal of the North American Benthological Society 21: 384–396.

Mati BM, Mutie S, Gadain H, Home P, Mtalo F. 2008. Impacts of land-use/ cover changes on the hydrology of the transboundary Mara River, Kenya/ Tanzania. Lakes and Reservoirs: Research and Management 13: 169–177.

Matiru V. 2000. *Forest cover in Kenya, policy and practice.* IUCN-World Conservation Union: Nairobi.

MATLAB and Statistics Toolbox Release 2013b, The MathWorks, Inc., Natick, Massachusetts, United States.

May F, Ash J. 1990. An assessment of the allelopathic potential of Eucalyptus. Aust. J. Bot. 38, 245.

Mbabazi D. 2004. Trophic characterization of the dominant fishes in the Victoria and Kyoga Lake Basins. PhD dissertation. Makerere University, Uganda.

McCartney BA. 2010.. Evaluation of water quality and aquatic ecosystem health in the Mara River basin, East Africa. MSc thesis, Florida International University, Miami, Florida.

McClain ME, Subalusky AL, Anderson EP, Dessu SB, Melesse AM, Ndomba PM, Mtamba JOD, Tamatamah RA, Mligo C. 2014. Comparing flow regime, channel hydraulics and biological communities to infer flow–ecology relationships in the Mara River of Kenya and Tanzania. Hydrological Sciences Journal 59: 1–19.

McClain ME. 2013. Balancing Water Resources Development and Environmental Sustainability in Africa: A Review of Recent Research Findings and Applications. Ambio 42: 549-565.

McCutchan JH, Lewis WM Jr, Kendall C, McGrath CC. 2003. Variation in trophic shift for stable isotope ratios of carbon, nitrogen, and sulfur. Oikos 102: 378–390.

McKnight D, Boyer E, Westerhoff P, Doran P, Kulbe T, Andersen D. 2001. Characterization of Dissolved Organic Matter for Indication of Precursor Organic Material and Aromaticity Limnology and Oceanography 46:38-48

McTammany ME, Benfield EF, Webster JR. 2007. Recovery of stream ecosystem metabolism from historical agriculture. Journal of the North American Benthological Society 26: 532-545.

Mead LH, Wiegner TN. 2010. Surface water metabolism in a tropical estuary, Hilo Bay, Hawaii, USA, during storm and non-storm conditions. Estuaries and Coasts 33: 1099–1112.

Melesse A, McClain M, Abira M, Mutayoba W, Wang XM. 2008. Modeling the Impact of Land-Cover and Rainfall Regime Change Scenarios on the Flow of Mara River, Kenya ASCE-EWRI. World Environmental and Water Resources Congress, DOI 10.1061/40976(316)558.

Merritt RW, Cummins KW, Berg MB, Novak JA, Higgins MJ, Wessell KJ, Lessard JL. 2002. Development and application of a macroinvertebrate functional-group approach to the bioassessment of remnant river oxbows in southwest Florida. Journal of the North American Benthological Society 21: 290–310.

Merritt RW, Cummins KW, Berg MB. (eds.) 2008. An introduction to the Aquatic Insects of North America. Kendall / Hunt Publishing Company, Dubuque, Iowa.

Merritt RW, Cummins KW. 1996. An introduction to the aquatic insects of North America. Kendall/Hunt: Dubuque, Iowa.

Merritt RW, Cummins KW. 2006. Trophic relationships of macroinvertebrates. Pages 585–610 in F. R. Hauer and G. A. Lamberti (editors). Methods in stream ecology. 2nd edition. Academic Press, San Diego, California.

Meyer JL, Edwards TR. 1990. Ecosystem metabolism and turnover of organic carbon along a blackwater river continuum. Ecology 71: 668–677.

Meyer JL, Wallace JB, Eggert SL. 1998. Leaf litter as a source of dissolved organic carbon in streams. Ecosystems 1: 240-249.

Meyer JL. 1989. Can P/R ratio be used to assess the food base of stream ecosystems: A comment on Rosenfeld and Mackay (1987). Oikos 54: 119–121.

Minagawa M, Wada E. 1984. Stepwise enrichment of ^{15}N along food chains: further evidence and the relation between δ^{15}N and animal age. Geochimica et Cosmochimica Acta 48: 1135–1140.

Minaya V, McClain ME, Moog O, Omengo F, Singer G. 2013. Scale-dependent effects of rural activities on enthic macroinvertebrates and physico-chemical characteristics in headwater streams of the Mara River, Kenya. Ecological Indicators 32:116–122.

Minshall GW, Cummins KW, Petersen RC, Cushing CE, Bruns DA., Sedell J., Vannote R L. 1985. Developments in stream ecosystem theory. Canadian Journal of Fisheries and Aquatic Sciences 42: 1045-1055.

Miserendino ML, Pizzolon LA. 2004. Interactive effects of basin features and land-use change on macroinvertebrate communities of headwater streams in the Patagonian Andes. River Research Applications 20: 967–983.

Morara GO, Mathooko JM, Leichtfried M. 2003. Natural leaf litter transport and retention in a second-order tropical stream: the Njoro River, Kenya. African Journal of Ecology 41: 277-279.

Morrongiello JR, Bond NR, Crook DA, Wong BBM. 2011. Eucalyptus leachate inhibits reproduction in a freshwater fish. Freshwater Biology 56: 1736–1745.

Mosepele K, Moyle PB, Merron GS, Purkey DR, Mosepele B. 2009. Fish, Floods, and Ecosystem Engineers: Aquatic Conservation in the Okavango Delta, Botswana. BioScience 59: 53–64.

Moss B. 2007. Rapid shredding of leaves by crabs in a tropical African stream. Verhandlungen der Internationalen Vereinigung für theoretische und angewandte Limnologie 29:147–150.

Moulton TP, Magalhães-Fraga SAP, Brito EF and Barbosa FA. 2010. Macroconsumers are more important than specialist macroinvertebrate shredders in leaf processing in urban forest streams of Rio de Janeiro, Brazil. Hydrobiologia 638: 55–66.

Moulton TP, Souza ML, Silveira RML, Krsulović FAM. 2004. Effects of ephemeropterans and shrimps on periphyton and sediments in a stream in Atlantic forest, Rio de Janeiro, Brazil. Journal of the North American Benthological Society 23: 868–881.

Mpawenayo B, Mathooko JM. 2005. The structure of diatom assemblages associated with Cladophora and sediments in a highland stream in Kenya. Hydrobiologia 544: 55–67.

Mugidde R, Hecky RE, Hendzel L, Taylor WD. 2003. Pelagic nitrogen fixation in Lake Victoria, Uganda. Journal of Great Lakes Research 29: 76–88.

Mugo J, Tweddle D. 1999. Preliminary surveys of the fish and fisheries of the Nzoia, Nyando and Sondu/Miriu rivers, Kenya. Part I. In Report of Third FIDAWOG Workshop (Tweddle, D. & Cowx, I. G., eds), pp. 106–125. LVFRP Technical Report 99/06.

Mulholland PJ and others. 2001. Inter-biome comparison of factors controlling stream metabolism. Freshwater Biology 46: 1503–1517, doi:10.1046/j.1365-2427.2001.00773.x

Mulholland PJ, Tank JL, Sanzone DM, Wollheim WM, Peterson BJ, Webster JR, Meyer JL. 2000. Nitrogen cycling in a forest stream determined by a ^{15}N tracer addition. Ecological Monographs, 70: 471-493.

Mulholland PJ, Marzolf ER, Webster JR, Hart DR, Hendricks SP. 1997. Evidence that hyporheic zones increase heterotrophic metabolism and phosphorus uptake in forest streams. Limnology and Oceanography 42: 443-451.

Munishi PKT. 2007. The biodiversity values of the Mara River (Masurura) Swamp, Mara region, northern Tanzania. WWF-TZ Report, Dar es Salaam, Tanzania.

Murphy JF, Giller PS. 2000. Seasonal dynamics of macroinvertebrate assemblages in the benthos and associated with detritus packs in two low-order streams with different riparian vegetation. Freshwater Biology 43: 617–631.

Muyodi FJ, Bugenyi FWB, Hecky RE. 2010. Experiences and lessons learned from interventions in the Lake Victoria Basin: The Lake Victoria Environmental Management Project. Lakes & Reservoirs: Research and Management 15: 77–88.

Muyodi FJ, Hecky RE, Kitamirike JM, Odong R. 2009. Trends in health risks from water-related diseases and cyanotoxins in Ugandan portion of Lake Victoria basin. Lakes & Reservoirs: Research and Management 14: 247–257.

Mwamburi J. 2003. Variations in trace elements in bottom sediments of major rivers in Lake Victoria's basin, Kenya. Lakes & Reservoirs: Research and Management 252: 5-13.

Mwashote BM, Shimbira W. 1994. Some limnological characteristics of the lower Sondu-Miriu River, Kenya. In: *Okemwa, E.; Wakwabi, E.O.; Getabu, A. (Ed.) Proceedings of the Second EEC Regional Seminar on Recent Trends of Research on Lake Victoria Fisheries, Nairobi : ICIPE Science Press, p. 15-27.*

Naiman RJ, Alldredge JR, Beauchamp DA, Bisson PA, Congleton J, Henny CJ, Huntly N, Lamberson R, Levings R, Merrill EN, Pearcy WG, Rieman BE, Ruggerone GT, Scarnecchia D, Smouse PE and Wood CC. 2012. Developing a broader scientific foundation for river restoration: Columbia River food webs. PNAS109 (52): 21201–21207.

Naiman RJ, Braak L, Grant R, Kemp AC, du Toit JT, Venter FJ. 2003. Interactions between species and ecosystem characteristics. In: du Toit J, Biggs H, Rogers KH. (Eds), The Kruger experience: ecology and management of Savanna heterogeneity, 221–241 pp. Washington, DC: Island Press.

Naiman RJ, Decamps H, McClain ME. 2005. Riparia: Ecology, Conservation, and Management of Streamside Communities. Elsevier Academic Press.

Naiman RJ, Décamps H, Pastor J, Johnston CA. 1988. The potential importance of boundaries to fluvial ecosystems. Journal of the North American Benthological Society 7: 289-306.

Naiman RJ, Melillo JM, Hobbie JE. 1986. Ecosystem alteration of boreal forest streams by beaver (*Castor canadensis*). Ecology 67:1254-1269.

Naiman RJ, Rogers KH. 1997. Large animals and system level characteristics in river corridors. BioScience 47: 521-529.

Nakano S, Miyasaka H, Kuhara N. 1999. Terrestrial-aquatic linkages: riparian arthropod inputs alter trophic cascades in a stream food web. Ecology 80: 2435-2441.

NBI (Nile Basin Initiative) 2005. Nile TransboundaryEnvironment Action Project. Status of water quality monitoring in the Kenyan portion of Lake Victoria Basin. Khartoum, Sudan.

NBI (Nile Basin Initiative). 2009. Nile Transboundary Environment Action Project. Sio-Siteko Transboundary Wetland Community Management Plan. Khartoum, Sudan.

NBI (Nile Basin Initiative). 2011. Kenya and the Nile basin initiative; Benefits of cooperation, NBI Secretariat. Khartoum, Sudan.

Nel JL, Roux DJ, Abell R, Ashton PJ, Cowling RM, Higgins JV, Thieme M, Viers JH. 2009. Progress and challenges in freshwater conservation planning. Aquatic Conservation: Marine and Freshwater Ecosystems 19: 474-485.

Nicholson SE, Yin X, Ba MB. 2000. On the feasibility of using a lake water balance model to infer rainfall: an example from Lake Victoria. Hydrological Sciences Journal 45: 75-95.

Nilsson C, Reidy CA, Dynesius M, Revenga C. 2005. Fragmentation and flow regulation of the world's large river systems. Science 308: 405-408.

Njiru M, Okeyo-Owuor JB, Muchiri M and Cowx IG. 2004. Shifts in the food of Nile tilapia, Oreochromis niloticus (L.) in Lake Victoria, Kenya. African Journal of Ecology 42: 163–170.

Njuguna PK. 1996. Building an inventory of Kenya's wetlands: a biological inventory of wetlands of Uashin Gishu District of Kenya. Kenya Wetlands Working Group: Nairobi.

Nyenje PM, Foppen JW, Uhlenbrook S, Kulabako R, Muwanga A. 2010. Eutrophication and nutrient release in urban areas of sub-Saharan Africa - A review. Science of the Total Environment 408: 447–455.

Obati GO. 2007. An Investigation of Forest Ecosystem Health in Relation to Anthropogenic Disturbance in the Southwestern Mau Forest Reserve, Kenya. PhD Thesis, University of Bremen, Germany.

Ochumba PBO, Kibaara DI. 1989. Observations on blue-green algal blooms in the open waters of Lake Victoria, Kenya. African Journal of Ecology 27: 23–34.

Ochumba PBO, Manyala JO. 1992. Distribution of fishes along the Sondu-Miriu River of Lake Victoria, Kenya with special reference to upstream migration, biology and yield. Aquaculture and Fish Management 23:701–719.

Odada EO, Ochola WO, Olago DO. 2009. Drivers of ecosystem change and their impacts on human well-being in Lake Victoria basin. African Journal of Ecology 47: 46–54.

Offula AVO, Karanja D, Omondi R, Okurut T, Matano A, Jembe T, Abila R, Boera P, Gichuki J. (2010). Relative abundance of mosquitoes and snails associated with water hyacinth and hippo grass in the Nyanza gulf of Lake Victoria. Lakes & Reservoirs: Research and Management 15: 255–271.

Ogle SM, Breidt FJ, Paustian K. 2005. Agricultural management impacts on soil organic carbon storage under moist and dry climatic conditions of temperate and tropical regions. Biogeochemistry 72: 87-121.

Ogutu JO, Owen-Smith N, Piepho HP, Said MY. 2011. Continuing wildlife population declines and range contraction in the Mara region of Kenya during 1977–2009. Journal of Zoology 285: 99–109.

Ogutu JO, Piepho H-P, Dublin HT, Bhola N, Reid RS. 2007. El Niño-Southern Oscillation, rainfall, temperature and Normalized Difference Vegetation Index fluctuations in the Mara-Serengeti ecosystem. African Journal Ecology 46: 132–143.

Ogutu JO, Piepho H-P, Reid RS, Rainy ME, Kruska RL, Worden JS, Nyabenge M, Hobbs NT. 2010. Large herbivore responses to water and settlements in savannas. Ecological Monographs 80: 241–266.

Ogutu-Ohwayo R. 1990. The decline of the native fishes of Lake Victoria and Kyoga (East Africa) and the impact of introduced species, especially the Nile perch, Lates niloticus and the Nile tilapia, *Oreochromis niloticus*. Environmental Biology of Fishes 27: 81–96.

Ojunga S, Masese FO, Manyala JO, Etiegni E, Onkware AO, Senelwa K, Raburu PO, Balozi BK, Omutange ES. 2010. Impact of a Kraft Pulp and Paper Mill Effluent on Phytoplankton and Macroinvertebrates in River Nzoia, Kenya. Water Quality Research Journal of Canada 45: 235-250.

Ojuok JE. 2005. *Distribution, status and some aspects of the biology of two non-cichlid native fishes of Lake Victoria, Kenya*. In Knowledge and Experiences Gained from Managing the Lake Victoria Ecosystem, Mallya GA, Katagira FF, Kang'oha G, Mbwana SB, Katunzi EF,Wambede JT, Azza N, Wakwabi E, Njoka SW, Kusewa M, Busulwa H (eds). Regional Secretariat, Lake Victoria Environmental Management Project (LVEMP): Dar es Salaam; 309–317 pp.

Ojwang WO, Kaufman L, Agembe S, Asila A. 2004. Isotopic evidence of functional overlap amongst the resilient pelagic fishes of Lake Victoria, Kenya. Hydrobiologia 529: 27-35.

Ojwang WO, Kaufman L, Soule E, Asila AA. 2007. Evidence of stenotopy and anthropogenic influence on carbon source for two major riverine fishes of the Lake Victoria watershed. Journal of Fish Biology 70: 1430–1446.

Ojwang WO, Ojuok JE, Mbabazi D, Kaufman L. 2010. Ubiquitous omnivory, functional redundancy and the resiliency of Lake Victoria fish community. Aquatic Ecosystem Health and Management 13: 269-276.

Ojwang WOO. 2006. Patterns of resurgence and anthropogenic influence on trophic sources and interactions among fishes of Lake Victoria, Kenya. Ph.D Thesis. Boston University, Boston.

Okedi J. 1971. The food and feeding habits of the small mormyrid fishes of Lake Victoria, East Africa. African Journal of Tropical Hydrobiology and Fisheries 1: 1–12.

Okungu J, Opango P. 2005. Pollution loads into Lake Victoria from the Kenyan catchment. In Knowledge and Experiences gained from Managing the Lake Victoria Ecosystem, Mallya GA, Katagira FF, Kang'oha G, Mbwana SB, Katunzi EF, Wambede JT, Azza N, Wakwabi E, Njoka SW, Kusewa M, Busulwa H (eds). Regional Secretariat, Lake Victoria Environmental Management Project (LVEMP): Dar es Salaam; 90-108.

Omengo FO. 2010. Carbon cycling in the Mara River system – Influence of land use and location in the fluvial network on organic matter processing and CO2-production. MSc thesis, UNESCO-IHE Institute for Water Education, Delft, the Netherlands.

Omondi R, Ogari J. 1994. *Preliminary study on the food and feeding habits of Schilbe mystus (Linn., 1762) in River Nyando*. In: Okemwa, E.; Wakwabi, E.O.; Getabu, A. (Ed.) Proceedings of the Second EEC Regional Seminar on Recent Trends of Research on Lake Victoria Fisheries. ICIPE Science: Nairobi, p. 115-119.

Ongeri DMK, Lalah JO, Wandiga SO, Schramm KW and Michalke EB. 2009. Levels of Toxic Metals in Multisectoral Samples from Winam Gulf of Lake Victoria. Bulletin of Environmental Contamination and Toxicology 82: 64–69.

Ortega-Retuerta E, Frazer TK, Duarte CM, Ruiz S, Tovar-Sánchez A, Arrieta JM, Reche I. 2009. Biogeneration of chromophoric dissolved organic matter by bacteria and krill in the Southern Ocean. Limnology & Oceanography 54: 1941–1950.

Ortiz-Zayas JR, Lewis WM Jr, Saunders JF III, McCutchan JH Jr, Scatena FN. 2005. Metabolism of a tropical rainforest stream. Journal of the North American Benthological Society 24: 769–783.

Osborne LL, Wiley MJ. 1992. Influence of tributary spatial position on the structure of warmwater fish communities. Canadian Journal of Fisheries and Aquatic Sciences 49: 671–681.

Oyoo-Okoth E, Admiraal W, Osano O, Ngure V,.Kraak MHS, Omutange ES. 2011. Monitoring exposure to heavy metals among children in LakeVictoria, Kenya: Environmental and fish matrix. Ecotoxicology and Environmental Safety 73:1797-1803, DOI:10.1016/j.ecoenv.2010.07.040.

Oyugi D. 2011. Freshwater Ecoregions of the World: Lake Victoria basin. WWF/TNC. http://www.feow.org/ecoregion_details.php?eco=521. Accessed 11/09/2011.

Paetzold A, Sabo JL, Sadler JP, Findlay SEG, Tockner K. 2007. Aquatic–terrestrial subsidies along river corridors. In: Wood PJ, Hannah DM, Sadler JP. (Eds), Hydroecology and Ecohydrology: Past, Present and Future, pp 57-92. John Wiley and Sons, Chichester, UK.

Palmer C, O'Keeffe J, Palmer A, Dunne T, Radloff S. 1993. Macroinvertebrate functional feeding groups in the middle and lower reaches the Buffalo River, Eastern Cape, South Africa. I. Dietary variability. Freshwater Biology 29:441–453.

Palmer MA. Filoso S. 2009. Restoration of ecosystem services for environmental markets. Science 325: 575–576.

Parlanti E, Wörz K, Geoffroy L, Lamotte M. 2000. Dissolved organic matter fluorescence spectroscopy as a tool to estimate biological activity in a coastal zone submitted to anthropogenic inputs. Organic Geochemistry 31: 1765-1781.

Parnell AC, Inger R, Bearhop S, Jackson AL. 2010. Source partitioning using stable isotopes: coping with too much variation. PlosOne 5: e9672.

Pellika P, Lötjönen M, Siljander M, Lens L. 2009. Airborne remote sensing of spatiotemporal change (1955-2004) in indigenous exotic forest cover in the Taita Hills, Kenya. International Journal of Applied Earth Observation and Geoinformation 11: 221–232.

Petrone KC, Richards JS, Grierson PF. 2009. Bioavailability and composition of dissolved organic carbon and nitrogen in a near coastal catchment of south-western Australia. Biogeochemistry 92:27–40.

Peuravuori J, Pihlaja K. 1997. Molecular size distribution and spectroscopic properties of aquatic humic substances. Analytica Chimica Acta 337: 133-149.

Phillips DL, Gregg JW. 2003. Source partitioning using stable isotopes: coping with too many sources. Oecologia 136: 261–269.

Pingram MA, Collier KJ, Hamilton DP, Hicks BJ, David BO. 2014. Spatial and temporal patterns of carbon flow in a temperate, large river food web. Hydrobiologia 729: 107–131.

Pinnegar JK, Polunin NVC. 1999. Differential fractionation of $\delta^{13}C$ and $\delta^{15}N$ among fish tissue: implication for the study of trophic interactions. Functional Ecology 13: 225–231.

Poff NL, Allan JD, Bain MB, Karr JR, Prestegaard KL, Richter BD, Sparks RE, Stromberg JC. 1997. The natural flow regime: a paradigm for river conservation and restoration. BioScience 47: 769–784.

Poff NL, Olden JD, Pepin DM, Bledsoe BP. 2006. Placing global streamflow variability in geographic and geomorphic context. River Research and Applications 22: 149-166.

Poff NL, Zimmerman JKH. 2010. Ecological responses to altered flow regimes: a literature review to inform the science and management of environmental flows. Freshwater Biology 55: 194–205.

Polis GA, Anderson WB, Holt RD. 1997. Toward an integration of landscape and food web ecology: The dynamics of spatially subsidized food webs. Annual Review of Ecology and Systematics 28: 289-316.

Post AWCHM. 2008. The hippopotamus: nothing but a nuisance? Hippo-human conflicts in Lake Victoria area, Kenya. MSc thesis, University of Amsterdam, Amsterdam.

Post DM, Layman CA, Arrington DA, Takimoto G, Quattrochi J, Montana CG. 2007. Getting to the fat of the matter: models, methods and assumptions for dealing with lipids in stable isotope analyses. Oecologia 152: 179–189.

Post DM, Taylor JP, Kitchell JF, Olson MH, Schindler DE, Herwig BR. 1998. The role of migratory waterfowl as nutrient vectors in a managed wetland. Conservation Biology 12: 910-920.

Post DM. 2002. Using stable isotopes to estimate trophic position: Models, methods, and assumptions. Ecology 83: 703-718.

Pozo J, Casas J, Menéndez M, Mollá S, Arostegui I, Basaguren A, Salinas MJ. 2011. Leaf-litter decomposition in headwater streams: a comparison of the process among four climatic regions. Journal of the North American Benthological Society 30: 935–950.

Pringle CM, Hamazaki T. 1998. The role of omnivory in a neotropical stream: separating diurnal and nocturnal effects. Ecology 79: 269–280.

Prins HHT. 2000. Competition between wildlife and livestock in Africa. In: Prins HHT, Grootenhuis JG, Dolan TT. (Eds), Wildlife conservation by sustainable use: 51–80. Boston, MA: Kluwer Academic Press.

Pusey BJ, Arthington AH, Stewart-Koster B, Kennard MJ, Read MG. 2010. Widespread omnivory and low temporal and spatial variation in the diet of fishes in a hydrologically variable northern Australian river. Journal of Fish Biology 77: 731–753.

Pusey BJ, Read MG, Arthington AH. 1995. The feeding ecology of freshwater fishes in two rivers of the Australian Wet Tropics. Environmental Biology of Fishes 43: 85–103.

Quinton JN, Govers G, Van Oost K, Bardgett RD. 2010. The impact of agricultural soil erosion on biogeochemical cycling. Nature Geosciences 3(5): 311-314.

Raburu PO, Masese FO. 2012. A fish-based index for assessing ecological integrity of riverine ecosystems in Lake Victoria Basin. River Research and Applications 28: 23–38.

Raburu PO, Okeyo-Owuor JB, Masese FO. 2009. Macroinvertebrate-based Index of Biotic Integrity (M-IBI) for Monitoring the Nyando River, Lake Victoria Basin, Kenya. Scientific Research and Essays 4: 1468-1477.

Raburu PO. 2003. Water quality and the status of aquatic macroinvertebrates and ichthyofauna in River Nyando, Kenya. PhD thesis, Moi University, Kenya.

Rasmussen JB, Trudeau V, Morinville GR. 2009. Estimating the scale of fish feeding movements in rivers using $\delta^{13}C$ signature gradients. Journal of Animal Ecology 78: 674–685.

Rasmussen JJ, Baattrup-Pedersen A, Riis T, Friberg N. 2011. Stream ecosystem properties and processes along a temperature gradient. Aquatic Ecology 45: 231–242.

Rathbun RE, Stephens DW, Shultz DJ, Tai DY. 1978. Laboratory studies of gas tracers for reaeration: American Society of Civil Engineers. Journal of Environmental Engineering Division 104, no. EE2: 215-229.

Rayner TS, Pusey BP, Pearson RG. 2009. Spatio-temporal dynamics of fish feeding in the lower Mulgrave River, north-eastern Queensland: the influence of seasonal flooding, instream productivity and invertebrate abundance. Marine and Freshwater Research 60: 97–111.

Reid RS, Rainy M, Ogutu J, Kruska RL, Kimani K, Nyabenge M, McCartney M, Kshatriya M, Worden J, Ng'ang'a L, Owuor J, Kinoti J, Njuguna E, Wilson CJ, Lamprey R. 2003. *People, Wildlife and Livestock in the Mara Ecosystem: the Mara Count 2002*. Report, Mara Count 2003, International Livestock Research Institute, Nairobi, Kenya.

Riley AJ, Dodds WK. 2013. Whole-stream metabolism: strategies for measuring and modeling diel trends of dissolved oxygen. Freshwater Science 32: 56–69.

Roach KA. 2013. Environmental factors affecting incorporation of terrestrial material into large river food webs. Freshwater Science 32(1): 283-298.

Roberts B, Mulholland P, Hill W. 2007. Multiple Scales of Temporal Variability in Ecosystem Metabolism Rates: Results from 2 Years of Continuous Monitoring in a Forested Headwater Stream. Ecosystems 10: 588-606.

Rodhe H, Virji H. 1976. Trends and periodicities in East African rainfall data. Montly Weather Review 104: 307–315.

Rosemond AD, Pringle CM, Ramírez A, Paul MJ, Meyer JL. 2002. Landscape variation in phosphorus concentration and effects on detritus-based tropical streams. Limnology & Oceanography 47: 278–289.

Rosenberg DM, Resh VH. (eds.) 1993. Freshwater Biomonitoring and Benthic Macroinvertebrates. Chapman Hall, New York

Rosenfeld JS, Mackay RJ. 1987. Assessing the food base of stream ecosystems: Alternatives to the P/R ratio. Oikos 50: 141–147.

Salmah MRC, Al-Shami SA, Hassan AA, Madrus MR, Huda AN. 2013. Distribution of detritivores in tropical forest streams of peninsular Malaysia: role of temperature, canopy cover and altitude variability. International journal of biometeorology: 1-12.

Sanderman J, Lohse KA, Baldock JA, Amundson R. 2009. Linking soils and streams: sources and chemistry of dissolved organic matter in a small coastal watershed, Water Resources Research 45: W034018.

Sangale F, Okungu J, Opango P. 2005. Variation of flow of water from Rivers Nzoia, Yala and Sio into Lake Victoria. In: Knowledge and Experiences gained from Managing the Lake Victoria Ecosystem, Mallya GA, Katagira FF, Kang'oha G, Mbwana SB, Katunzi EF,Wambede JT, Azza N,Wakwabi E, Njoka SW, Kusewa M, Busulwa H (eds), 691-702pp. Regional Secretariat, Lake Victoria Environmental Management Project (LVEMP): Dar es Salaam.

Scarsbrook MR, Halliday J. 1999. Transition from pasture to native forest land-use along stream continua: effects on stream ecosystems and implications for restoration. New Zealand Journal of Marine and Freshwater 33: 293–310.

Scheren PAGM, Zanting HA, Lemmens AMC. 2000. Estimation of water pollution sources in Lake Victoria, East Africa: Application and elaboration of the rapid assessment methodology. Journal of Environmental Management 58: 235-248.

Schlünz B, Schneider RR. 2000. Transport of terrestrial organic carbon to the oceans by rivers: re-estimating flux-and burial rates. International Journal of Earth Sciences 88: 599-606.

Schmitz OJ, Hawlena D, Trussell GC. 2010. Predator control of ecosystem nutrient dynamics. Ecology Letters 13: 1199-1209.

Seitzinger SP, Sanders RW (1997). Contribution of dissolved organic nitrogen from rivers to estuarine eutrophication. Marine Ecological Progress Series 159: 1–12.

Serneels S, Said MY, Lambin EF. 2001. Land-cover changes around a major East African wildlife reserve: the Mara Ecosystem (Kenya). International Journal of Remote Sensing 22: 3397-3420.

Shafroth PB, Wilcox AC, Lytle DA, Hickey JT, Andersen DC, Beauchamp VB, Hautzinger A, McMullen LE, Warner A. 2010. Ecosystem effects of environmental flows: modelling and experimental floods in a dryland river. Freshwater Biology 55: 68–85.

Shepherd K, Walsh M, Mugo F, Ong C, Hansen TS, Swallow B, Awiti A, Hai M, Nyantika B, Ombao D, Grunder M, Mbote F, Mungai D. 2000. Improved Land Management in the Lake Victoria Basin: Linking Land and Lake, Research and Extension, Catchment and Lake Basin. Final Technical Report. International Centre for Research in Agro-Forestry, Nairobi.

Shivoga WA. 2001. The influence of hydrology on the structure of invertebrate communities in two streams flowing into Lake Nakuru, Kenya. Hydrobiologia 458: 121–130.

SigmaPlot for Windows Version 12.0. Systat Software Inc., San Jose, California.

Silsibe GM, Hecky RE, Guildford SJ, Muggidde R. 2006. Variability of chlorophyll a and photosynthetic parameters in a nutrient-saturated tropical great lake. Limnology and Oceanography 51: 2052–2063.

Silva-Junior EF, Moulton TP, Boëchat IG, Gücker B. 2014. Leaf decomposition and ecosystem metabolism as functional indicators of land use impacts on tropical streams. Ecological Indicators 36: 195-204.

Sitoki L, Gichuki J, Ezekiel C, Wanda F, Mkumbo OC and Marshall BE. 2010. The Environment of Lake Victoria (East Africa): Current Status and Historical Changes. International Review of Hydrobiology 95: 209–223.

Smart JS. 1972. Quantitative characterization of channel network structure. Water Resources Research 8: 1487-1496.

Smith WE, Kwak TJ. 2014. Otolith microchemistry of tropical diadromous fishes: spatial and migratory dynamics. Journal of fish biology 84: 913-928.

Sombroek WG, Braun HMH, Van Der Pouw BJA. 1982. The explanatory soil map and agro-climatic zone map of Kenya. Report no. E.I, Kenya Soil Survey, Nairobi, Kenya.

Spencer RGM, Bolton L, Baker A. 2007. Freeze/thaw and pH effects on freshwater dissolved organic matter fluorescence and absorbance properties from a number of UK locations. Water Research 41: 2941–50.

Staehr P, Testa J, Kemp W, Cole J, Sand-Jensen K, Smith S. 2012. The metabolism of aquatic ecosystems: history, applications, and future challenges. Aquatic Sciences - Research Across Boundaries 74: 15-29.

Stager JC, Hecky RE, Grzesik D, Cumming BF, Kling H. 2009. Diatom evidence for the timing and causes of eutrophication in Lake Victoria, East Africa. Hydrobiologia 636: 463–478.

Stals, R., de Moor, I.J., 2007. Guides to the Freshwater Invertebrates of Southern Africa, Volume 10: Coleoptera. WRC report No. TT 320/07, South Africa

STATISTICA for Windows Version 7. StatSoft Incorporated, Tulsa, Oklahoma.

Stone MK, Wallace JB (1998) Long–term recovery of a mountain stream from clear–cut logging: the effects of forest succession on benthic invertebrate community structure. Freshwater Biology 39:151–169.

Strauch AM. 2011. Seasonal variability in faecal bacteria of semiarid rivers in the Serengeti National Park, Tanzania. Marine and Freshwater Research 62: 1191-1200.

Strauch AM. 2013. Interactions between soil, rainfall, and wildlife drive surface water quality across a savanna ecosystem. Ecohydrology 6: 94-103.

Subalusky AL, Dutton CL, Rosi-Marshall EJ, Post DM. 2014. The hippopotamus conveyor belt: vectors of carbon and nutrients from terrestrial grasslands to aquatic systems in sub-Saharan Africa. Freshwater Biology, doi:10.1111/fwb.12474.

Sutcliffe JV Parks YP. 1999. The Hydrology of the Nile. IAHS Special Publ. 5, IAHS Press: Wallingford, UK.

Sutcliffe JV, Petersen G. 2007. Lake Victoria: Derivation of a corrected natural water level series. Hydrological Sciences Journal 52: 1316-1321.

Sutherland BA, Meyer JL, Gardiner EP. 2002. Effects of land cover on sediment regime and fish assemblage structure in four southern Appalachian streams. Freshwater Biology 47: 1791–1805.

Talling JF. 1966. The annual cycle of stratification and phytoplankton growth in Lake Victoria (East Africa). Internationale Revue der gesamten Hydrobiologie 51: 545–621.

Tank JL, Rosi-Marshall EJ, Griffiths NA, Entrekin SA, Stephen ML. 2010. A review of allochthonous organic matter dynamics and metabolism in streams. Journal of the North American Benthological Society 29: 118-146.

Tate E, Sutcliffe J, Conway D, Farquharson F. 2004. Water balance of Lake Victoria: Update to 2000 and climate change modelling to 2100. Hydrological Sciences Journal 49: 563–574.

Thorp JH, Delong AD. 2002. Dominance of autochthonous autotrophic carbon in food webs of heterotrophic rivers. Oikos 96: 543–550.

Thorp JH, Delong MD. 1994. The riverine productivity model: an heuristic view of carbon sources and organic processing in large river ecosystems. Oikos 70: 305–308.

Thorp JH, Thoms MC, Delong MD. 2006. The riverine ecosystem synthesis: Biocomplexity in river networks across space and time. River Research and Applications 22: 123-147.

Tieszen LL, Seyimba MM, Imbamba SK, Troughton JH. 1979. The distribution of C3 and C4 grasses and carbon isotope discrimination along a altitudinal and moisture gradient in Kenya. Oecologia 37: 337–350.

Tockner K, Malard F, Ward JV. 2000. An extension of the flood pulse concept. Hydrological Processes 14: 2861-2883.

Townsend SA, Webster IT, Schult JH. 2011. Metabolism in a groundwater-fed river system in the Australian wet/dry tropics: tight coupling of photosynthesis and respiration. Journal of the North American Benthological Society 30: 603-620.

Tsivoglou BC, Neal LA. 1976. Tracer measurement of reaeration. HI. Predicting the capacity of inland streams. Journal of the Water Pollution Control Federation48: 2669-2689.

Tumwesigye C, Yususf SK, Makanga B. 2000. Structure and composition of benthic macroinvertebrates of a tropical forest stream, River Nyamweru, Western Uganda. African Journal of Ecology 38: 72-77.

Twesigye CK, Onywere SM, Getenga ZM, Mwakalila S, Nakiranda JK. 2011. The Impact of Land Use Activities on Vegetation Cover and Water Quality in the Lake Victoria Watershed. The Open Environmental Engineering Journal 4: 66-77.

Vander Zanden MJ, Rasmussen JB. 1999. Primary consumer delta C-13 and delta N-15 and the trophic position of aquatic consumers. Ecology 80: 1395-1404.

Vanni MJ. 2002. Nutrient cycling by animals in freshwater ecosystems. Annual Review of Ecology and Systematics 33:341-370.

Vannote RL, Minshall GW, Cummins KW, Sedell JR, Cushing CE. 1980. The river continuum concept. Canadian Journal of Fisheries and Aquatic Sciences 37: 130-137.

Verschuren D, Johnson TC, Kling HJ, Edgington DN, Leavitt PR, Brown ET, Talbot MR, Hecky RE 2002. History and timing of human impact on Lake Victoria, East Africa. Proceedings of the Royal Society of London B. 269: 289-294.

Vörösmarty CJ, McIntyre PB, Gessner MO, Dudgeon D, Prusevich A, Green P, Glidden S.et al. 2010. Global threats to human water security and river biodiversity. Nature 467: 555-561.

Wallace JB, Eggert SL, Meyer JL, Webster JR. 1997. Multiple trophic levels of a forest stream linked to terrestrial litter inputs. Science 277: 102–104.

Wallace JB, Webster JR. 1996. The role of macroinvertebrates in stream ecosystem function. Annual Review of Entomology 41: 115–139.

Walters AW, Barnes RT, Post DM. 2009. Anadromous alewives (Alosa pseudoharengus) contribute marine-derived nutrients to coastal stream food webs. Canadian Journal of Fisheries and Aquatic Sciences 66:439-448.

Wandiga SO. 1981. The concentration of zinc, copper, lead, manganese, nickel and fluoride in rivers and lakes of Kenya. SINET: Ethiopian Journal of Science 3: 1.

Wan L, Lyons J. 2003. Fish and benthic macroinvertebrate assemblage as indicators of stream degradation in urbanizing watersheds. In: Simon, T.P. (ed.), Biological response signatures: indicator patterns using aquatic communities. CRC Press, Boca Raton, pp 113–120.

Wanink JH. 1998. The pelagic cyprinid Rastrineobola argentea as a crucial link in the disrupted ecosystem of Lake Victoria. Dwarf and Giants– African Adventures. Ponsen and Looijen: Wageningen, the Netherlands.

Wantzen KM, Yule CM, Mathooko JM, Pringle CM. 2008. Organic matter processing in tropical streams. In: Dudgeon D. (ed.), Tropical Stream Ecology, 44–65pp. Academic Press, London.

Wantzen, K. M., and R. Wagner. 2006. Detritus processing by invertebrate shredders: a neotropical–temperate comparison. Journal of the North American Benthological Society 25:216–232.

Wardle DA, Bardgett RD, Callaway RM, Van der Putten WH. 2011. Terrestrial Ecosystem Responses to Species Gains and Losses. Science 332: 1273-1277.

Wass P (ed.). 1995. Kenya's indigenous forests: status management and conservation. IUCN (Forest Conservation Programme) Gland and Cambridge

Wasswa J, Kimerire BT, Nkedi-Kizza P, Mbabazi J, Ssebugere P. 2011. Organochlorine pesticide residues in sediments from the Uganda side of Lake Victoria. Chemosphere 82: 130–136.

Weishaar JL, Aiken GR, Bergamaschi BA, Fram MS, Fujii R, Mopper K. 2003. Evaluation of specific ultraviolet absorbance as an indicator of the chemical composition and reactivity of dissolved organic carbon. Environmental Science & Technology 37: 4702-4708.

Welcome RL, Ryder RA, Sedell JA. 1989. Dynamics of fish assemblages in river systems- a synthesis. In: Dodge DP. (ed.), Proceedings of the International Large River Symposium.

Wetzel RG, Likens GE. 2000. Limnological analysis. 3rd edition. Springer-Verlag, New York.

Wetzel RG. 1992. Gradient-dominated ecosystems - sources and regulatory functions of dissolved organic matter in freshwater ecosystems. Hydrobiologia 229: 181–98.

Whitehead PJP. 1959a. The anadromous fishes of Lake Victoria. Revue de Zoologieetde Botanique Africaines 59: 329-363.

Wilcock RJ, Scarsbrook MR, Costley KJ, Nagels JW. 2002. Controlled release experiments to determine the effects of shade and plants on nutrient retention in a lowland stream. Hydrobiologia 485: 153–162.

Wiley MJ, Osborne LL, Larimore RW. 1990. Longitudinal structure of an agricultural prairie river system and its relationship to current stream ecosystem theory. Canadian Journal of Fisheries and Aquatic Sciences 47:373–384.

Williams CJ, Yamashita Y, Wilson HF, Jaffe R, Xenopoulos MA. 2010. Unraveling the role of land use and microbial activity in shaping dissolved organic matter characteristics in stream ecosystems, Limnology & Oceanography 55: 1159-1171.

Williamson CE, Dodds W, Kratz TK, Palmer MA. 2008. Lakes and streams as sentinels of environmental change in terrestrial and atmospheric processes, Frontiers in Ecology and the Environment 6: 247-254.

Wilson HF, Xenopoulos MA. 2009. Effects of agricultural land use on the composition of fluvial dissolved organic matter. Nat Geosci 2:37–41.

Winemiller KO, Jepsen DB. 1998. Effects of seasonality and fish movement on tropical river food webs. Journal of Fish Biology 53: 267-296.

Winemiller KO, Kelso-Winemiller C. 2003. Food habits of tilapiine cichlids of the Upper Zambezi River and floodplain during the descending phase of the hydrologic cycle. Journal of Fish Biology 63: 120–128.

Winemiller KO. 1990. Spatial and temporal variation in tropical fish trophic networks. Ecological Monographs 60: 331–364.

Winemiller KO. 2004. Floodplain river food webs: generalizations and implications for fisheries management. In: Welcomme RL, Petr T. (Eds), Proceedings of the Second International Symposium on the Management of Large Rivers for Fisheries. Volume 2 Pages 285–312. Mekong River Commission, Phnom Penh, Cambodia.

Winterbourn MJ, Rounick JR, Cowie B. 1981. Are New Zealand stream ecosystems really different? New Zealand journal of marine and freshwater research 15: 321-328.

Wipfli MS, Hudson J, Caouette J. 1998. Influence of salmon carcasses on stream productivity: response of biofilm and benthic macroinvertebrates in southeastern Alaska, USA. Canadian Journal of Fisheries and Aquatic Sciences 55:1503-1511.

Wipfli MS, Richardson JS, Naiman RJ. 2007. Ecological Linkages between headwaters and downstream ecosystems: transport of organic matter, invertebrates, and wood down headwater channels. Journal of the American Water Resources Association 43(1): 72-85.

Wipfli MW, Baxter CV. 2010. Linking ecosystems, food webs, and fish production: subsidies in salmonid watersheds. Fisheries 35: 373–387.

Witte F, Goldschmidt T, Gouldswaard PC, Ligtvoet W, Van Oijen MJP, Wanink JH. 1992. Species introduction and concomitant ecological changes in Lake Victoria. Netherlands Journal of Zoology 42: 214–232.

Wolanski E, Gereta E. 1999. Oxygen cycle in a hippo pool, Serengeti National Park, Tanzania. African Journal of Ecology 37: 419–423.

Woodward G, Gessner MO, Giller PS, Gulis V, Hladyz S, Lecerf A, Malmqvist B, Mckie BG, Tiegs SD, Cariss H, Dobson M, Elosegi A, Ferreira V, Graca MAS, Fleituch T, Lacoursiere JO, Nistorescu M, Pozo J, Risnoveanu G, Schindler M, Vadineanu A, Vought LBM, Chauvet E. 2012. Continental scale effects of nutrient pollution on stream ecosystem functioning. Science 336: 1438–1440.

Woodward G, Hildrew AG. 2002. Food web structure in riverine landscapes. Freshwater Biology 47: 777–798.

Worthington EB. 1933. Inland waters of Africa. The results of two expeditions on the great Lakes of Kenya and Uganda, with accounts of their biology, native tribes and development. MacMillan and Co.: London.

WQBAR (Water Quality Baseline Assessment Report) 2007. Mara River Basin, Kenya-Tanzania Global Water for Sustainability Program, Florida International University, 61p.

WREM. 2008. Mara River Basin Monograph. Water Resources and Energy Management (WREM) International Inc.

Yates AG, Brua RB, Culp JM, Chambers PA. 2012. Multi-scaled drivers of rural prairie stream metabolism along human activity gradients. Freshwater Biology 58: 675-689.

Yillia PT, Kreuzinger N, Mathooko JM. 2008. The effect of in-stream activities on the Njoro River, Kenya. Part II: Microbial water quality. Physics and Chemistry of the Earth 33: 729-737.

Young RG, Huryn AD. 1998. Comment: Improvements to the diurnal upstream-downstream dissolved oxygen change technique for determining whole-stream metabolism in small streams. Canadian Journal of Fisheries and Aquatic Sciences 55: 1784-1785.

Young RG, Huryn AD. 1999. Effects of land use on stream metabolism and organic matter turnover, Ecological Applications 9(4): 1359-1376.

Young RG, Matthaei CD, Townsend CR. 2008. Organic matter breakdown and ecosystem metabolism: functional indicators for assessing river ecosystem health. Journal of the North American Benthological Society 27: 605–25.

Yule CM, Leong MY, Liew KC, Ratnarajah L, Schmidt K, Wong HM, Pearson RG, Boyero L. 2009. Shredders in Malaysia: abundance and richness are higher in cool upland tropical streams. Journal of the North American Benthological Society 28: 404–415.

Yule CM. 1996. Trophic relationships and food webs of the benthic invertebrate fauna of two aseasonal tropical streams on Bougainville Island, Papua New Guinea. Journal of Tropical Ecology 12: 517–534.

Zalewski M. 2002. Ecohydrology- the use of ecological and hydrological processes for sustainable management of water resources. Hydrological Sciences Journal 47: 825-834.

Zar JH. 1999. Biostatistical analysis. 4th edition. Prentice Hall, Upper Saddle River, NJ.

Zeug SC, Winemiller KO. 2008. Evidence supporting the importance of terrestrial carbon in a large-river food web. Ecology 89: 1733–1743.

Zimov SA. 2005. Pleistocene park: return of the mammoth's ecosystem. Science 308:796–798.

Samenvatting

Om rivieren en hun ecosysteemdiensten te behouden, herstellen en beheren is kennis over hun functionele dynamiek onontbeerlijk. Er is echter nog steeds een gebrek aan kennis over het functioneren van tropische rivieren met betrekking tot de energetische basis van riviervisserij. In deze studie wordt een overzicht gepresenteerd van de antropogene invloeden op omzettingen van organisch materiaal, energiebronnen en kenmerken van voedselketens in rivieren in het stroomgebied van het Victoriameer, met daarbij ook recente onderzoeksresultaten uit de tropen. Er zijn tegenstrijdige resultaten gerapporteerd over de diversiteit van versnipperaars en hun rol in het omzetten van organisch materiaal in tropische rivieren. Recent tropisch onderzoek heeft ook het belang van autochtoon koolstof, zelfs in kleine rivieren in beboste gebieden, benadrukt. Vergelijkbare onderzoeken in Afro-tropische rivieren zijn echter beperkt, wat het moeilijk maakt om hun positie in optredende patronen van koolstof stromen in de tropen vast te stellen.

Deze studie werd uitgevoerd in de Mara rivier, een belangrijke grensoverschrijdende rivier die ontspringt in het Mau boscomplex in Kenia en via Tanzania uitmondt in het Victoriameer. Bovenstrooms wordt het stroomgebied afgewaterd door twee zijrivieren, de Amala en Nyangores rivieren welke samenkomen in het middengedeelte en zo de hoofdstroom van de Mara vormen. De algemene doelstelling van deze dissertatie was het verkrijgen van een beter begrip over het functioneren van de Mara rivier door het bepalen van de dynamiek in ruimte en tijd van oorsprong en aanvoer van organische stof onder verschillende omstandigheden van landgebruik en stromingspatronen, en de invloed die deze dynamiek heeft op de energiestroom voor consumenten in de rivier. Benthische macro-invertebraten werden verzameld in rivieren met open en gesloten bladerdak en werden geclassificeerd naar functionele voedingsgroep op basis van een analyse van hun maaginhoud. In totaal werden 43 predatoren, 26 verzamelaars, 19 schrapers and 19 versnipperaars geïdentificeerd. De soortenrijkdom was hoger in rivieren met een gesloten bladerdak, waar versnipperaars ook de meest dominante groep waren in termen van biomassa. Zeven taxa van versnipperaars kwamen alléén voor in rivieren met een gesloten bladerdak, hetgeen het belang aangeeft van het behoud van water- en habitat kwaliteit, inclusief de toevoer van bladafval van de juiste kwaliteit, in de bestudeerde rivieren. De resultaten suggereren dat rivieren in hooglanden een diverse groep van versnipperaars herbergen in tegenstelling tot eerdere onderzoeken die slechts een beperkt aantal taxa van versnipperaars hebben aangetoond.

Vervolgens werden de samenstelling van invertebrate functionele voedingsgroepen en het ecosysteemproces van bladafbraak gebruikt als, respectievelijk, structurele en functionele indicator voor de conditie van het ecosysteem in Keniaanse hoogland rivieren. Grof- en fijnmazige strooiselzakken werden gebruikt om microbiële (fijnmazig) en versnipperaar + microbiële (grofmazig) afbraaksnelheden te vergelijken, en daarnaast voor het bepalen van de rol van versnipperaars in strooiselomzetting van bladeren van verschillende boomsoorten (inheemse *Croton macrostachyus* en *Syzygium cordatum* en de exoot *Eucalyptus globulus*). Voor de inheemse bladsoorten waren afbraaksnelheden over het algemeen

189

hoger in grofmazige dan in fijnmazige strooiselzakken en de relatieve verschillen in afbraaksnelheden tussen bladsoorten bleven ongewijzigd in zowel rivieren in landbouwgebieden als in bossen. Versnipperaars waren relatief belangrijker in bosrivieren dan in rivieren in landbouwgebieden, waar microbiële afbraak belangrijker was. Bovendien was de door versnipperaars gefaciliteerde bladafbraak afhankelijk van bladsoort, en deze was het hoogst voor *C. macrostachyus* en het laagst voor *E. globulus*, hetgeen suggereert dat vervanging van inheemse oevervegetatie door kwalitatief mindere *Eucalyptus* soorten langs rivieren tot een reductie in nutrienten recycling in rivieren kan leiden.

Om de dynamiek van organische stof in deze rivieren te bestuderen werd de invloed van verandering in landgebruik op de samenstelling en concentratie van opgelost organisch materiaal (OOM) bepaald en de relatie met rivier-ecosysteem metabolisme onderzocht. Optische eigenschappen van OOM gaven opmerkelijke verschuivingen in samenstelling te zien langs een gradient van landgebruik. Bosrivieren bevatten OOM met een hoger moleculair gewicht, afkomstig van land, terwijl rivieren in landbouwgebieden in verband gebracht werden met autochtoon geproduceerd, laagmolecuair OOM, en met fotodegradatie dankzij het open bladerdak. Geur was echter sterk op alle lokaties, onafhankelijk van landgebruik in het betreffende stroomgebied. In landbouw gebieden was een sterke geur waarschijnlijk afkomstig van landbouwgrond waar bodems worden gemobiliseerd tijdens het bewerken en vervolgens door run-off in rivieren terecht komen. Bruto primaire productie en ecosysteem respiratie waren over het algemeen hoger in rivieren in landbouwgebieden, vanwege enigszins open bladerdak en hogere nutrientenconcentraties. De conclusies van deze studie zijn belangrijk omdat ze, naast het bevestigen van de rol van tropische rivieren in de mondiale koolstofcyclus, ze ook de gevolgen aantonen van veranderingen in landgebruik op het functioneren van ecosystemen in een regio waar landgebruik zal intensiveren als gevolg van populatiegroei.

Tot slot werden natuurlijke dichtheden van stabiele koolstof (δ^{13}C) en stikstof (δ^{15}N) isotopen gebruikt om patronen van koolstofstromen in voedselketens langs de longitudinale gradient van de Mara rivier in ruimte en tijd te kwantificeren. Stukken rivier werden geselecteerd die onder verschillende invloed stonden van herbivoren (vee en wild). Potentiële primaire producenten (terrestrische C3 en C4 producenten en perifyton) en consumenten (invertebraten en vis) werden verzameld tijdens droge en natte seizoenen die zo een reeks van contrasterende stroomcondities vertegenwoordigden. "Stable Isotope Analysis in R (SIAR) Bayesian mixing modellering" werd gebruikt om terrestrische en autochtone bronnen van organische koolstof voor consumenten te scheiden. In het algemeen domineerde de bijdrage van perifyton tijdens het droge seizoen. Gedurende het natte seizoen was het belang van terrestrisch koolstof hoger met een afnemend belang van C3-producenten met toenemende afstand van de beboste bovenstroomse gebieden terwijl het belang van C4-producenten toenam in riviergebieden die input ontvingen van vee en nijlpaarden. Deze studie benadrukt het belang van grote herbivore zoogdieren op het functioneren van rivierecosystemen en de gevolgen van het verdwijnen van savannelandschappen die momenteel de nog resterende populaties van deze dieren herbergen.

De resultaten van deze dissertatie dragen bij aan discussies over de effecten van veranderingen in landgebruik van hoogland rivieren en voedselketens in savanne rivieren op koolstof stromen en de rol die grote herbivore zoogdieren spelen wanneer ze terrestrisch organisch materiaal in rivieren verplaatsen. Deze studie verschaft ook informatie en aanbevelingen voor toekomstig onderzoek en beheersactiviteiten die moeten leiden tot duurzaamheid van de Mara rivier en gerelateerde ecosystemen in het stroomgebied van het Victoriameer.

About the author

Mr. Frank Onderi Masese was born in Bo'Masisi, Kisii County, Kenya. He is a holder of BSc in Fisheries (Hons, 2004) and MSc in Aquatic Sciences (2008) from Moi University, Kenya. Mr. Masese has worked and participated in a number of research projects and consultancies in western Kenya, including development of management programmes for wetlands, development of biomonitoring programmes for riverine ecosystems and peace building initiatives in resource-use conflict areas. Before the PhD fellowship, Mr. Masese was working as a Research Associate with Victoria Institute for Research on Environment and Development (VIRED) International, an NGO based in western Kenya, that deals with environmental conservation and promoting wise use of wetland resources. In addition he was participating in a project "Methods for Valuation, Attribution and Compensation for Environmental Services in Eastern and Central Africa". He also worked as a part-time lecturer at Moi University, Eldoret, Kenya.

Mr. Masese's PhD programme was a sandwich arrangement whereby he conducted his field work in Kenya in conjunction with Egerton University and course work, consultation and writing of the thesis took place at UNESCO-IHE. During the PhD tenure he also guided seven MSc students during proposal development, field work in the Mara and thesis writing. Mr. Masese is a member of various professional organisations; Ecological Society of Eastern Africa (ESEA), East African Water Association (EAWA), Nature Kenya, and the Ecological Society of America (ESA).

Journal publications

1. **Masese FO**, Salcedo-Borda JS, Gettel GM, Irvine K, McClain ME. Linkage between metabolism and DOM composition in headwater streams influenced by different land use. In Submission
2. **Masese FO**, Abrantes KG, Gettel GM, Bouillon S, Irvine K, McClain ME. Are large herbivores vectors of terrestrial subsidies for riverine food webs? *Ecosystems,* in press.
3. **Masese FO**, Kitaka N, Kipkemboi J, Gettel GM, Irvine K, McClain ME. 2014. Litter processing and shredder distribution as indicators of riparian and catchment influences on ecological health of tropical streams. *Ecological Indicators* 46: 23–37.
4. **Masese FO**, Kitaka N, Kipkemboi J, Gettel GM, Irvine K, McClain ME. 2014. Macroinvertebrate functional feeding groups in Kenyan highland streams: more evidence for a diverse shredder assemblage. *Freshwater Science* 33: 435-450.
5. Kilonzo F, **Masese FO**, Van Griensven A, Bauwens W, Obando J, Lens PNL. 2014. Spatial-temporal variability in water quality and macro-invertebrate assemblages in the Upper Mara River basin, Kenya. *Physics and Chemistry of the Earth* 67–69: 93–104.
6. Gichana Z. M, Njiru M., Raburu, P.O. and **Masese F.O.** (2014). Effects of human activities on microbial water quality of the Nyangores stream, Mare River basin, IJSTR 3(2).
7. **Masese FO**, Omukoto JO, Kobingi Nyakeya K (2013). Biomonitoring as a pre-requisite for sustainable water resources: a review of current status and challenges in East Africa. *Ecohydrology & Hydrobiology* 13 (3): 173–191.
8. **Masese FO** and McClain ME (2012). Trophic sources and emergent food web attributes in rivers of the Lake Victoria Basin: a review with reference to anthropogenic influences. *Ecohydrology* 5: 685–707.
9. Raburu PO and **Masese FO** (2012). A fish-based index for assessing ecological integrity of riverine ecosystems in Lake Victoria Basin. *River Research and Applications* 28: 23–38.
10. Ojunga S, **Masese F**O, Manyala JO, Etiegni L, Onkware AO, Senelwa K, Raburu PO, Balozi BK, Omutange ES (2010). Impact of a Kraft Pulp and Paper Mill Effluent on Phytoplankton and Macroinvertebrates in River Nzoia, Kenya. *Water Quality Research Journal of Canada* 45: 235-250
11. **Masese FO**, Raburu PO and Muchiri M. (2009). A preliminary benthic macroinvertebrates index of biotic integrity (B-IBI) for monitoring the Moiben River, Lake Victoria Basin, Kenya. *African Journal of Aquatic Science.* 34: 1-14.
12. **Masese FO**, Muchiri M. and Raburu PO. (2009). Macroinvertebrate assemblages as biological indicators of water quality in the Moiben River, Kenya. *African Journal of Aquatic Science* 34: 15-26.

13. Kobingi N, Raburu PO, **Masese FO** and Gichuki J. (2009). Assessment of pollution impacts on ecological Integrity of the Kisian and Kisat Rivers in Lake Victoria Drainage Basin, Kenya. *African Journal of Environmental Science and Technology* 3: 97-107.

14. Raburu PO, JB Okeyo-Owuor and **Masese FO** (2009). Macroinvertebrate-based Index of Biotic Integrity (M-IBI) for Monitoring the Nyando River, Lake Victoria Basin, Kenya. *Scientific Research and Essays* 4: 1468-1477.

15. Raburu PO, **Masese FO** and Mulanda CA. (2009). Macroinvertebrate Index of Biotic Integrity (M-IBI) for monitoring rivers in the upper catchment of Lake Victoria Basin, Kenya. *Aquatic Ecosystem Health and Management* 12: 197-205.

Contribution to book chapters

1. Njiru M, Sitoki L, Nyamweya CS, Jembe J, Aura CM, Waithaka E and **Masese FO** (2012). Environmental degradation and ecological changes in the Lake Victoria fisheries: causes, issues and management. NOVA Publishers, New York, pp 1-27.

2. Okeyo-Owuor J.B, Raburu P.O., **Masese F.O.** and Omari S.N. (2012). Wetlands of Lake Victoria Basin, Kenya: distribution, current status and conservation challenges. In: *Community Based Approaches to the Management of Nyando Wetlands, Lake Victoria Basin, Kenya*, PO Raburu, Okeyo-Owuor JB, Kwena F, (Eds), 1-14 pp. KDC–VIRED-UNDP, Nairobi.

3. Raburu P.O., Khisa P. and **Masese F.O.** (2012). Background Information on Nyando Wetland. In: *Community Based Approaches to the Management of Nyando Wetlands, Lake Victoria Basin, Kenya*, PO Raburu, Okeyo-Owuor JB, Kwena F, (Eds), 15-31 pp. KDC–VIRED-UNDP, Nairobi.

4. **Masese F.O**, Raburu P.O and Kwena F. (2012). Threats to the Nyando Wetland. In: *Community Based Approaches to the Management of Nyando Wetlands, Lake Victoria Basin, Kenya*, PO Raburu, Okeyo-Owuor JB, Kwena F, (Eds), 68-80 pp. KDC–VIRED-UNDP, Nairobi.

5. Obiero K.O, Raburu P.O and **Masese F.O.** (2012). Managing Nyando Wetland in the face of Climate Change. In: *Community Based Approaches to the Management of Nyando Wetlands, Lake Victoria Basin, Kenya*, PO Raburu, Okeyo-Owuor JB, Kwena F, (Eds), 100-121 pp. KDC–VIRED-UNDP, Nairobi.

6. Opaa B.O., Okotto-Okotto J., Nyandiga C.O. and **Masese, F.O.** (2012). Nyando Wetland in the Future. In: *Community Based Approaches to the Management of Nyando Wetlands, Lake Victoria Basin, Kenya*, PO Raburu, Okeyo-Owuor JB, Kwena F, (Eds), 100-121 pp. KDC–VIRED-UNDP, Nairobi.

7. **Masese FO**, Raburu PO, Mwasi BN and Etiégni L (2011). Effects of Deforestation on Water Resources: Integrating Science and Community Perspectives in the Sondu-Miriu River Basin, Kenya. In: *New Advances and Contributions to Forestry Research*, Oteng-Amoako AA (ed.), 1-18 pp. InTech Open Access Publisher, Riejka.

8. Okeyo-Owuor JB, **FO Mases**e, H Mogaka, E Okwuosa, G Kairu, P Nantongo, A Agasha, & B Biryahwaho (2011). Status, Challenges and New Approaches for Management of the Trans-Boundary Mt. Elgon Ecosystem: a Review. In: *Towards Implementation of Payment for Environmental Services (PES): a collection of findings linked to the ASARECA funded research activities,* 60-82 pp. VDM Verlag Dr. Müller, Saarbrücken, 404p.

Conferences/ workshop presentations

Masese FO, Gettel GM, Irvine K, McClain ME. (2013). Downstream shifts in the basal energy sources of the Mara River, Kenya. NCR-Days 2013 Conference: Teaming up with nature and nations, 3rd-4th October, 2013, Delft, the Netherlands

Masese FO (2013). Downstream shifts in the basal energy sources of the Mara River, Kenya. Invited presentation, Antwerp University, Belgium

Masese FO, Gettel GM, Irvine K, McClain ME. (2013). Dynamics in trophic sources in riverine food webs: effects of flow variation and subsidies. UNESCO-IHE PhD Seminar, September 2013, Delft, The Netherlands.

Masese FO, Irvine K, Gettel G, McClain M. (2013). Macroinvertebrate structural and functional responses to changes in hydraulic and physical conditions in the Mara River, Kenya. 4th International Multidsciplinary Conference on Hydrology and Ecology: Emerging Patterns, Breakthroughs and Challenges, 13th-16th May 2013, Rennes, France.

Masese FO, McClain ME, Kitaka N, Kipkemboi J, Gettel GM, Irvine K. (2012). Decomposition of leaf litter in agricultural and forest streams in the Upper Mara River basin, Kenya. The 13th WaterNet/WARFSA/GWP-SA Symposium, October 2012, Johannesburg, South Africa.

Masese FO, McClain ME, Kitaka N, Kipkemboi J, Gettel GM, Irvine K. (2012). Macroinvertebrate shredders in upland Kenyan streams; influence of catchment land use on abundance, biomass and distribution: The 3rd LVBC Scientific Conference on the Lake Victoria Basin "Harnessing Research for Sustainable Development of the Lake Victoria Basin" 22nd to 23rd October, 2012, Entebbe, Uganda.

Masese FO (2012). Land use influences on the decomposition of leaf litter in upland streams in Kenya. The annual SENSE PhD seminar, University of Twente, The Netherlands.

Masese FO (2011). An analysis of the relative roles of shredders and microbes in decomposition of leaf litter in agricultural and forest streams in the Upper Mara River basin, Kenya. UNESCO-IHE Seminar, Delft, The Netherlands.

Masese FO (2010). Biomonitoring of ecological integrity of riverine ecosystems in the Lake Victoria basin, Kenya: status, potential benefits and challenges to scaling up. Aquatic Resources of Kenya (ARK II): Aquatic Research for Development, 2nd International Conference.

Masese FO (2010). Spatio-temporal dynamics in the sources, processing and supply of organic matter and influences on food web architecture in the tropical Mara River. UNESCO-IHE PhD Seminar, Delft, The Netherlands.

Masese FO, Raburu PO (2009). Moi University 5th International Conference: Tolerance of Baetidae, Caenidae and Hydropsychidae to disturbance: Implications for EPT Index performance and a search for alternatives in the Lake Victoria Basin, Kenya.

Netherlands Research School for the
Socio-Economic and Natural Sciences of the Environment

DIPLOMA

For specialised PhD training

The Netherlands Research School for the
Socio-Economic and Natural Sciences of the Environment
(SENSE) declares that

Frank Onderi Masese

born on 17 January 1980 in Kisii, Kenya

has successfully fulfilled all requirements of the
Educational Programme of SENSE.

Delft, 11 March 2015

the Chairman of the SENSE board   the SENSE Director of Education

Prof. dr. Huub Rijnaarts   Dr. Ad van Dommelen

The SENSE Research School has been accredited by the Royal Netherlands Academy of Arts and Sciences (KNAW)

KONINKLIJKE NEDERLANDSE
AKADEMIE VAN WETENSCHAPPEN

The SENSE Research School declares that Mr Frank O. Masese has successfully fulfilled all requirements of the Educational PhD Programme of SENSE with a work load of 42.4 EC, including the following activities:

SENSE PhD Courses

o   Environmental Research in Context (2010)
o   Research in Context Activity: Organising Workshop on 'Integrating research outcomes with water resources management in the Mara River basin', Kenya (2014)

Other PhD and Advanced MSc Courses

o   Environmental Flows Assessment, UNESCO-IHE Delft (2010)
o   Measurement of Stream Ecosystem Metabolism, UNESCO-IHE, Kenya (2011)
o   Environmental Data Quality, University of Twente (2012)
o   Flow Systems Analysis, UNESCO-IHE Delft (2012)

Management and Didactic Skills Training

o   Supervision of two MSc thesis students, Moi University, Kenya (2011-2012)
o   Supervision of two MSc thesis students, Egerton University, Kenya (2012)
o   Supervision of two MSc thesis students, UNESCO-IHE Delft, The Netherlands (2012-2014)
o   Teaching MSc course 'Stream and river ecology module' - Joint Limnology and Wetlands Management Programme, Egerton University, Kenya (2012-2014)

Oral Presentations

o   *Large herbivore as vectors of terrestrial subsidies for riverine food webs.* 1st Annual International Interdisciplinary Conference: Africa and the New World Order, 30 July-2 August 2014, Kisii, Kenya
o   *Downstream shifts in the basal energy sources of the Mara River, Kenya.* Netherlands Centre for River Studies (NCR)-Days Conference: Teaming up with nature and nations, 3-4 October 2013, Delft, The Netherlands
o   *Decomposition of leaf litter in agricultural and forest streams in the upper Mara River basin, Kenya.* 13th WaterNet / WARFSA / GWP-SA Symposium, 31 October-2 November, Johannesburg, South Africa
o   *Macroinvertebrate shredders in upland Kenyan streams; influence of catchment land use on abundance, biomass and distribution.* 3rd Scientific Conference on the Lake Victoria Basin (LVBC), 23 October 2012, Entebbe, Uganda

SENSE Coordinator PhD Education

Dr. ing. Monique Gulickx

Printed and bound by CPI Group (UK) Ltd, Croydon, CR0 4YY

21/10/2024

01777101-0010